T0178788

Small-scale rural biogas programmes

Praise for this book

'This book provides an excellent practical overview of biogas extension programmes, discussing challenges faced in different parts of the world, and the initiation and management of implementation programmes. It provides a clear overview of how anaerobic digestion actually works, focusing on design aspects that will help the reader to select a digester design, set it up and keep it going. It deals with the equipment needed to use the gas, and the important issue of how to use the valuable bioslurry residue produced by the digestion process. The book is thorough and well written from a practical perspective. I highly recommend it to anyone interested in practical implementation of small-scale digesters in rural areas.'

*Jo Smith, Professor of soil organic matter and nutrient
modelling, University of Aberdeen, UK*

'David Fulford's extensive knowledge of biogas through decades of work in the UK and particularly overseas has been brought into this excellent and timely book which is concerned with technical aspects of biogas use in rural communities.'

*Maria M. Vahdati is a lecturer in Renewable Energy
at the University of Reading, UK.*

'Anyone involved in biogas work will treasure this book because it provides details of the science of biogas, implementation and management of biogas programmes. David Fulford once again helps us think more clearly about important issues in rural biogas implementation.'

*Vianney Tumwesige of Green Heat (U) Ltd,
Kampala, Uganda*

'*Small-scale Rural Biogas Programmes* is an excellent update in the field of biogas, which is a potential source of fuel around us. It is a useful reading material for Students of Environmental Science, and a reliable reference for those beginning and running biogas programmes.'

*Ainea Kimaro won the Ashden Awards 2005 and is the Managing Director
for Biogas and Solar Co. Ltd, Bagamoyo Tanzania*

Small-scale rural biogas programmes
A Handbook

David Fulford

PRACTICAL ACTION
Publishing

Practical Action Publishing Ltd
The Schumacher Centre
Bourton on Dunsmore, Rugby,
Warwickshire CV23 9QZ, UK
www.practicalactionpublishing.org

ISBN 978-1-85339-849-0 Hardback
ISBN 978-1-85339-850-6 Paperback
ISBN 978-1-78044-849-7 Library Ebook
ISBN 978-1-78044-850-3 Ebook

A catalogue record for this book is available from the British Library.

The author has asserted his rights under the Copyright Designs and
Patents Act 1988 to be identified as author of this work.

Fulford, D. (2015) *Small-scale, rural biogas programmes*,
Rugby, UK: Practical Action Publishing
<http://dx.doi.org/10.3362/9781780448497>

Since 1974, Practical Action Publishing has published and disseminated
books and information in support of international development work
throughout the world. Practical Action Publishing is a trading name
of Practical Action Publishing Ltd (Company Reg. No. 1159018), the
wholly owned publishing company of Practical Action. Practical Action
Publishing trades only in support of its parent charity objectives and
any profits are covenanted back to Practical Action (Charity Reg. No.
247257, Group VAT Registration No. 880 9924 76).

Cover photo: Nepali woman cooking with biogas.
Credit: David Fulford
Cover design by Mercer Design
Indexed by Liz Fawcett
Typeset by Allzone Digital Services Limited
Printed by Replika Press

FSC

Contents

http://dx.doi.org/10.3362/9781780448497.000

List of Figures

Figures

About the author

Dr David Fulford is in a unique position to write this book on biogas extension programmes. He was involved in the national biogas programme in Nepal, as part of the biogas team in Developing and Consulting Services (DCS) of United Mission to Nepal (UMN) under Sanfred Ruohoniemi. He and the rest of the team (Andrew Bulmer, John Finlay and Mamie Lau Wong) wrote an extended report for USAID, which was published by UMN (Biogas: Challenges and Experience from Nepal). He rewrote this report as a book (*Running a Biogas Programme: A handbook*), published by Practical Action Publishing (then Intermediate Technology Publishing) in 1988. After doing a PhD, he taught on an MSc course on renewable energy at the University of Reading. After leaving in 2007, he set up Kingdom Bioenergy Ltd (www.kingdombio.com) to do consulting work.

One consulting contract was as an assessor for Ashden (www.ashden.org), which enabled him to visit many renewable energy projects around the world, including several biogas extension programmes. He led a team that did a mid-term evaluation of the UNDP funded project in Nepal in 1990 that followed the one started under UMN and was on a team that evaluated the Nepal and Asia Biogas Programme run by SNV (Netherlands Development Organisation) in 2012.

David Fulford is a founding trustee of Foundation SKG Sangha (www. foundationskgsangha.org), a UK registered charitable company, set up with Vidya Sagar Devabhaktuni and Kiran Kumar Kudaravalli of SKG Sangha in Kolar, South India, to encourage biogas extension work.

Acknowledgements

I would like to thank those who have checked the manuscript of the book. My wife, Jane, checked and improved my grammar. Dr Marty Climenhaga of Southampton University checked several chapters and offered useful improvements to Chapter 4. Dr John Sathiaseelan Daniel, an agricultural consultant, checked Chapter 5. Acknowledgements are given to Govinda Devkota, who provided a copy of his book, which was published in Nepal and gives details of the technology used by BSP in Nepal.

The photographs in the book were all taken by me, although some were taken as part of Ashden assessment visits. The figures were all drawn by me. Some are based on dimensions taken from other publications.

Acronyms

ABS	Acrylonitrile Butadiene Styrene (a type of plastic)
ADB	Asian Development Bank
ADB/N	Agriculture Development Bank, Nepal
ADBL	Agriculture Development Bank Limited (Nepal)
ADM1	Advanced Digestion Model 1
AEPC	Alternative Energy Promotion Centre (Nepal)
AFPRO	Action for Food Production (Indian NGO)
ARTI	Appropriate Rural Technology Institute (Pune, Indian NGO)
BARC	Bhabha Atomic Research Centre (Mumbai, India)
BARD	Bangladesh Academy for Rural Development
BARI	Bangladesh Agricultural Research Institute
BCC	Biogas Coordinating Committee
BCSIR	Bangladesh Council of Scientific and Industrial Research
BOD	Biological Oxygen Demand
BORDA	Bremen Overseas Research and Development Association
BSP	Biogas Support Programme (Nepal)
BSP	British Standard Pipe thread
BSP/N	Biogas Sector Partnership, Nepal (NGO)
C:N	Carbon to Nitrogen ratio
CAMARTEC	Centre for Agricultural Mechanization and Rural Technology (Tanzania)
CDM	Clean Development Mechanism
CER	Certified Emissions Reduction
COD	Chemical Oxygen Demand
COF	Carbon Offset Finance
CSTR	Continuously Stirred Tank Reactor
DCS	Development and Consulting Services (Nepal)
DM	Dry Matter
DNES	Department of Non-conventional Energy Sources, India
DRANCO	DRy ANaerobic COmposting
EGSB	Expanded Granular Sludge Bed
EPA	Environmental Protection Agency (USA)
FAO	Food and Agriculture Organisation of the United Nations
FRP	Fibre (Glass) Reinforced Plastic
GATE	German Appropriate Technology Exchange or (Deutsches Zentrum für Entwicklungstechnologien)
GGC	Gobar Gas and Rural Equipment Development Company Ltd (Nepal)
GI	Galvanised iron (zinc coated steel)

GIZ	Gesellschaft für Internationale Zusammenarbeit (German Society for International Cooperation)
GPS	Global Positioning System
GS	Gold Standard
GS-VER	Gold Standard Voluntary Emissions Reduction
GTZ	Deutsche Gesellschaft für Technische Zusammenarbeit GmbH (now GIZ)
HDPE	High density polyethylene (a type of plastic)
HIVOS	Humanist Institute for Co-operation with Developing Countries (Netherlands)
HP	Horse Power (= 0.75 kW)
IBR	Induced Bed Reactor
ID	Inside diameter
IDCOL	Infrastructure Development Company, Bangladesh
IEA	International Energy Authority
ITDG	Intermediate Technology Development Group (now Practical Action)
KfW	Kreditanstalt für Wiederaufbau (German Development Bank)
KVIC	Khadi and Village Industries Commission, India
LCA	Life Cycle Analysis
LPG	Liquid Petroleum Gas
MARD	Ministry of Agriculture and Rural Development (Vietnam)
MFI	Micro-Finance Institute
MNES	Ministry of New and Renewable Energy Sources, India
MSW	Municipal Solid Waste
NABARD	National Bank for Agriculture and Rural Development (India)
NBMMP	National Biogas and Manure Management Programme (India)
NBPA	Nepal Biogas Promotion Association (Nepal)
NFFO	Non-Fossil Fuel Obligation (UK)
NGO	Non-Government Organization
NIAH	National Institute of Animal Husbandry (Vietnam)
NLSP	National Loan Subsidy Program (China)
NPBD	The National Project on Biogas Development (India)
NPT	National Pipe Thread
NTP	Normal temperature and pressure (usually 20 °C and 101 kPa)
OD	Outside diameter
OECD	Organisation for Economic Co-operation and Development
OPEC	Organisation of the Petroleum Exporting Countries
PoA-DD	Program of Activities Design Document
PPR	Polypropylene Random copolymer (a type of plastic)
PRAD	Planning Research and Action Division of the State Planning Institute of Uttar Pradesh, India
PTFE	Polytetrafluoroethylene (a type of plastic)
PVC	Polyvinyl Chloride (a type of plastic)
R&D	Research and Development

RBI	Reserve Bank of India
RET	Renewable Energy Technologies
SNV	Netherlands Development Organisation (Stichting Nederlandse Vrijwilligers)
TS	Total Solids
UASB	Upflow Anaerobic Sludge Blanket
UMN	United Mission to Nepal
UNCDF	United Nations Capital development Fund
UNDP	United Nations Development Programme
UNFCCC	United Nations Framework Convention on Climate Change
USAID	United States Agency for International Development
VACVINA	NGO encouraging VAC, a domestic garden system for vegetables (Vietnam)
VER	Voluntary Emissions Reduction
VFA	Volatile Fatty Acids
VGS	Voluntary Gold Standard
VS	Volatile Solids
WB	World Bank
WG	Water Gauge (the pressure produced by a column of water in mm)
WWF	World Wildlife Fund

Foreword

It gives me great pleasure to write the Foreword for this book. Sustainable energy, used at a local level, brings great social, environmental and economic benefits. And the use of biogas in rural homes exemplifies these benefits: as David Fulford says, it is a real 'win win win' technology.

So what are the benefits of biogas at a local level? Firstly, it uses waste materials to produce an energy-rich gas. This gas can replace wood for cooking, saving time and reducing the pressure on scarce forest resources. It can also replace fossil fuels for cooking or electricity generation, saving money.

Secondly, biogas reduces atmospheric pollution. Whether replacing wood or fossil fuel, biogas cuts greenhouse gas production, and it produces minimal local emissions. This is really important, since indoor air pollution is increasingly recognized as a huge global health threat, leading to over four million premature deaths each year. Much of that pollution arises from using wood for cooking, so changing to biogas can bring significant benefits to health – it can literally save lives.

Thirdly, biogas provides a way to manage organic waste, in particular manure, human sewage and wet food waste – materials which are messy, smell bad and spread disease. Management of these wastes is a growing challenge in many parts of the developing world, with more and more people moving to crowded towns and cities, and increasing emphasis on stall-fed livestock.

Finally, the residue from biogas can be a valuable soil improver, returning both nutrients and bulk to the soil. There is sometimes a misconception that putting manure into biogas conflicts with its important use as a fertilizer. But in fact the biogas plant neatly takes care of that, separating the energy-rich components from the nutrient-rich components of waste. Not a bad achievement for a tank of bacteria!

Great benefits – so what prevents biogas from being used more widely? Putting it simply, you cannot just plug a biogas plant in and switch it on – although it is a 'household' device it is more like a car than a refrigerator. So users need training on how to use biogas correctly, and regular checks and servicing are required to keep a plant in good condition. A great value of this book is that, along with technical detail, it also covers the organizational and social aspects of how to make biogas work as a programme or business. As with a car, the initial cost of a biogas plant is high, and different approaches that have been used to make biogas affordable are also discussed in this book.

I set up the charity Ashden in 2001 to champion the use of sustainable energy at a local level, because of the benefits it brings. David Fulford has made a great contribution to our work as an assessor for the Ashden Awards and particularly for the insight he has provided on biogas. I am delighted that through this book he is also bringing the achievements of Ashden winners to a wider audience.

This book should be the 'first port of call' for anyone new to the biogas sector. And with its detail and wide range of references, it is also a valuable information source for specialists already working in the sector.

Sarah Butler-Sloss, Founder Director of Ashden

CHAPTER 1
Overview of biogas extension

Abstract

An overview of biogas extension work considers the background to the development and use of the technology in less developed countries. The technology is placed in the wider global context of a concern for the environment since the early 1970s. It offers a long list of benefits to the users of the technology. A brief history looks at its origin in India and China, as well as applications for sewage processing in Europe and the USA. The benefits of biogas technology for small farmers led to programmes in China and India, which inspired the programme in Nepal. This book is the latest publication following a series of reports and a previous book about the biogas programme in Nepal. People in Europe and the USA have only lately realised that a technology for processing sewage is also a good source of renewable energy.

Keywords: biogas, anaerobic digestion, overview

The global context of the development of biogas extension programmes

Biogas has been seen as the classic example of a renewable energy technology or 'Appropriate Technology' since concerns were expressed at the rate of development in the world in the early 1970s. Several books at the time suggested that the world needed sources of energy other than oil. The key books were *Limits to Growth* (by Donella H. Meadows, Dennis L. Meadows, Jørgen Randers, and William W. Behrens III) (Meadows *et al.*, 1972) from the Club of Rome in 1972; '*Small is Beautiful: A Study of Economics as if People Mattered*' by E.F. Schumacher in 1973 (Schumacher, 1973); and 'Our Common Future', the report of the Brundtland Commission (set up in 1983, but published in 1987) (Brundtland, 1987). The two oil price-hikes of 1973–74 and 1979–80, caused first by OPEC restricting their supplies and then by the first Gulf war, confirmed this concern, especially for leaders in developing countries that were badly affected by the price hikes.

In India and China, various groups had been experimenting since the 1930s with using animal dung to generate biogas. The idea had been fairly well known before that time. There are reports of a sewage gas plant built in Bombay in 1859 and the idea was brought to the UK in 1895, when the gas produced was used to light street lamps. The 1970s oil price-hike encouraged both the Chinese and Indian governments to invest in programmes to spread the technology into rural areas. The second hike encouraged greater

http://dx.doi.org/10.3362/9781780448497.001

emphasis on biogas in India and China; it also encouraged governments to start programmes in many other countries across the world, particularly in the USA, the UK and in mainland Europe.

The rapid drop in oil prices in the mid-1980s meant that many biogas programmes faltered. Those in the USA, Europe and many other places were abandoned. In China and India, biogas technology was seen as a way of offering development to rural farmers, encouraging them to stay on the land rather than moving to the cities. These programmes therefore grew slowly but steadily from their initial foundations, which allowed a large amount of expertise and experience to be built up. The Chinese government, in particular, has been keen to share its expertise with other countries, including African nations. Various groups in Germany, especially Bremen Overseas Research and Development Association (BORDA) and German Appropriate Technology Exchange or Deutsches Zentrum für Entwicklungstechnologien (GATE) worked to spread information about these technologies in other parts of the world (Kossmann *et al.*, 1999; Sasse *et al.*, 1991).

The growing concern about the effects of climate change, combined with concerns about 'peak oil', since the early 1990s has triggered a new interest in biogas technology. The launch of the United Nations Framework Convention on Climate Change in 1994 (UN, 1992) and the subsequent Kyoto Protocol (UN, 2008) put pressure on many countries to develop alternatives to fossil fuels. Biogas is seen as an effective way to generate energy from agricultural and food residues that are often consigned as wastes for disposal.

The developed country that has taken most interest in biogas is Germany, based on its experience working in developing countries and also encouraged by the influential Green movement. Other European countries following close behind are Sweden, Denmark, Belgium and Austria. The USA has also developed a small, but steady, programme in biogas technology.

The countries that have built millions of biogas units, such as India and China, have a wealth of expertise that needs to be shared with the rest of the world.

Benefits of biogas technology

Biogas comes from anaerobic digestion, a process that uses naturally occurring microbes to break down food materials into methane and carbon dioxide in the absence of oxygen. The methane is very similar to natural gas, so is an easy-to-use high-grade fuel that can be used for cooking, heating or to run engines to generate shaft power or electricity. The presence of the carbon dioxide does reduce the heating value of the gas, but this does not seriously affect the value of biogas for these applications. However, the presence of carbon dioxide does make it more difficult to store biogas under pressure.

The microbes that break down food materials are similar to those used to compost the same materials. Aerobic composting generates water and carbon dioxide as well as heat. Using an anaerobic (without oxygen) process means the break down proceeds via a different pathway, so that energy is retained in the gas, rather than released as heat. The residue left after the digestion process is good compost, which can be used to enhance soil fertility and structure. The residue also has the huge benefit of having greatly reduced odour. Volatile fatty acids and other breakdown products are the source of the smells from rotting food and dung. Anaerobic microorganisms consume volatile fatty acids, which are the products of the breakdown of food materials by other microorganisms. Since methanogenic microorganisms use these volatile fatty acids to make biogas, they reduce the smell that the other microbes generate.

The microbes (which include both bacteria and archaea) that perform anaerobic digestion exist in the gut of cattle, as they help them digest grass and other plant foods. Most of the early programmes identified animal dung as the main feedstock material for the generation of biogas, because it already contains the right microbes. However, food material that has not been through the gut of an animal can also be used and produces much more biogas per kilogram, because the animal has not extracted a proportion of the food value for its own use.

In traditional farming systems, organic wastes are recycled back to the land, enabling soil fertility and structure to be conserved. The rapid growth in the use of inorganic fertilizer has increased crop yields, but at the cost of weakening soil structure. The centralization of food supply and waste disposal led to valuable organic matter being buried in landfill. This food waste rots in the anaerobic conditions in the landfill and releases methane, which is a potent greenhouse gas. While this gas can be collected and used to provide heat or generate electricity, the organic matter is permanently buried and lost to the topsoil. Biogas technology can be seen as a way of recycling both the energy and the organic matter available in food residues. The most effective systems are on-farm systems, as the compost made from the biogas plant effluent can be used directly on the fields nearby. This is one of the reasons that small-scale on-farm biogas systems have been so successful in China, India and Nepal.

For small-scale on-farm systems, the main use of the gas is for cooking food. In China, India and Nepal, a family can do all its cooking on the gas produced from dung produced by about four cattle or eight pigs (Ashden, 2005a). Biogas systems built for institutions, such as schools and hostels, in Kerala, India, have replaced 50 per cent of the LPG used for cooking with biogas produced from the sewage and food wastes that the institution produces (Ashden, 2007a). Prisons in Rwanda have replaced 50 per cent of the wood needed for cooking by using biogas from the sewage produced by the inmates (Ashden, 2005b). Stoves that burn biogas are fairly easy to design and make. Stoves for LPG or natural gas can be used with biogas, but only if they are properly adapted.

Biogas is a high-grade fuel that can be used in internal combustion engines. In the USA and Europe, the main use of biogas is to run engines to generate electricity. In Germany, Sweden and other countries in Europe, biogas is being cleaned of carbon dioxide, then pressurised and injected into the gas grid. Biogas has also been used to run vehicles and even a small train in Sweden. Since the source of the biogas is plant material that absorbs carbon dioxide from the atmosphere as it grows, the process is seen as carbon neutral. Biogas technology can therefore attract carbon credits and other subsidies that have been designed to encourage the replacement of fossil fuels with renewable sources of energy.

A brief history of biogas technology

The idea that a flammable gas can be generated from the break down of organic materials has been known for several centuries. Various sources (Harris, 2008) quote anecdotal evidence that suggest biogas from rotting wastes was used to heat bath water in Assyria in the 10th century BC. Other sources suggest that marsh gas was conveyed by bamboo pipes in China to provide energy in 400 BC (Hopkins, 2007). Also in China, there has been a history of covered cesspits, where the collection of night soil for compost has a very long tradition.

Various researchers in the past have investigated the flammable gas from rotting vegetation in marshes, although it was Alessandro Volta (Wolfe, 2004) in 1778, who identified this flammable gas as methane using marsh gas from Lake Maggiore. Humphrey Davey, found methane in 'fire-damp' in coal mines in 1815 (Hartley, 1960).

Anaerobic digestion has a strong, but fairly specialist, development history as the most effective way to process sewage. Cesspits, as a way of storing sewage, have been known since Roman times, but these are only useful if they are emptied regularly. As cities grew rapidly, especially in the 1800s, large concentrations of people made for large amounts of sewage, and in 1849 Dr John Snow (Frerichs, 2001) recognised that many diseases, such as typhoid and cholera, were spread through sewage. Attempts were made to install drains to direct the sewage away (Burian et al., 2000) or to collect it in cesspits (Alleman, 1982). Initially, these attempts meant that rivers flowing through urban areas became heavily polluted, but at least the sewage was washed away to the sea. Where there was a concern about the possibility of transfer of infection, cesspits were used to collect and isolate sewage from possible sources of disease.

Various sources (Harris, 2008) suggest that the first anaerobic digester was built at a leper colony in Bombay in 1859. It was probably built as a contained cesspit to prevent the spread of leprosy, but it was discovered that it generated a flammable gas. Research on biogas digestion was started in China in 1880s (Chen et al., 2010). Gas from an anaerobic digester was used to provide gas lights in 1897 at the Ackworth Leper Home Matunga in Bombay (Kansal et al., 1998). In 1907, the gas from this digester was used to run an internal combustion engine (Kurian, 2004). A Frenchman, John Louis Mouras (Cooper, 2001), patented the septic tank, in which the solid material is decomposed by anaerobic digestion. Dr Donald Cameron in Exeter, England built a version of the septic tank in 1895. Biogas was recovered from his 'Monster Septic Tank' and used to fuel street lamps (Enongene, 2003).

As engineers built sewage drainage systems in cities they discovered that the sewage generated biogas in the pipes, which rose to the highest points and blocked the pipes. The system in Sheffield in the UK had this problem, but vents at the highest places generated bad odours. In 1895, Joseph Edmund Webb, a builder from Birmingham, patented a 'sewer gas destructor', which was a gas street light that stayed on continuously to burn the sewage gas (Caldwell, 2008).

A sewage sludge digester was patented by Imhoff in Germany in 1906 (Gikas *et al.*, 2004; Parkin, 1986) and it separated the solid matter as sludge from the liquid wastes and then allowed the solid sludge to anaerobically digest. This was replaced with a two-stage system that used one chamber to separate the sludge from where it was transferred to a second chamber for digestion. In the 1920s, the two-stage sewage-sludge digester was further developed. In 1922, the gas from one such plant in Birmingham was used to power a gas engine that ran the pumps required in the operation of the plant (Hobson and Bousfield, 1981). In the 1920s and 1930s, active research by the Illinois State Water Survey Division in the USA provided a strong basis for the understanding of the different ways of treating sewage including the use of anaerobic digestion (Buswell, 1936, 1930). This work was continued at the University of Illinois.

Cylindrical sludge-digester tanks have been used very widely in Europe and in the USA since then and they generate biogas as the sludge is broken down. In many of these systems, the biogas has routinely been collected and used to heat the digester plant to speed the digestion process. During and after the Second World War, when petroleum was in short supply, electric generators were fitted to some sewage digesters, so that power could be generated using the gas. Some plant engineers ran their vehicles on pressurised gas as well, especially in the UK.

Many of these sewage-gas systems were seen as of peripheral interest until the early 1990s, when renewable energy became politically important. For example, many of the early projects funded by the Non-Fossil Fuel Obligation (NFFO) scheme in the UK were new sewage-gas-fuelled generators, replacing old systems that had become out of date. The approach to biogas technology in the USA and Europe has been based on large-scale centralized systems, using the experience gained with these sewage systems. In many places, particularly in parts of Europe, aerobic sewage-processing systems have been used in preference to anaerobic systems. However, aerobic systems require energy to run, whereas anaerobic systems generate energy in the form of biogas as they process the sewage.

The history of biogas technology in developing countries has been very different, as the technology has been small scale and geared to the needs of small farmers. The motive has been to replace firewood for cooking with an alternative clean energy source that also produced good compost. The main uptake has been in China, India and Nepal, where the programmes have been very successful and millions of biogas plants have been built.

There is a new interest across the world in the processing of food wastes using anaerobic digestion. As people move into cities from rural areas, food wastes cannot be composted and easily returned to the land. As cities become more developed, the gardens in which animals used to be kept have been built on, so the animals have stopped being available to eat food waste. People tend to add these food wastes to their waste bins, so they become part of the general solid wastes that must be removed and processed by municipal services. The same approach has been adopted by food processing industries, many of which have contracts with municipal services to dispose of their wastes. The general practice has been to put these wastes into landfill. Concern about both gaseous and liquid pollution from landfills has inspired regulators in many countries to impose requirements for the collection and burning off of the methane generated, either in flares, or in engines where energy can be recovered. The regulations are becoming even stricter in many countries prompting many groups to look at anaerobic digestion as a means to process food wastes.

Documentation of the biogas extension programme in Nepal and other countries in Asia

The biogas programmes in China and India developed from an initial interest in anaerobic digestion and act as models of how such programmes can function. The programme in Nepal was inspired by the experiences gained from these larger, nearby programmes. While the literature of biogas extension programmes is very patchy, especially for China and India, the programme in Nepal has been well documented from its beginning.

The programme in Nepal was launched in 1976 by the government of Nepal and the Development and Consulting Services (DCS) of The United Mission to Nepal (UMN) became involved and quickly developed their work into an effective national biogas programme. The original project, which was funded by United States Agency for International Development USAID, was the subject of a detailed report (Bulmer *et al.*, 1985), which was rewritten and published by Intermediate Technology as *Running a Biogas Programme* (Fulford, 1988). DCS set up the biogas extension work under a private company called Gobar Gas tatha Krishi Yantra Shala, (Biogas and Rural Equipment Company) Pvt Ltd, usually called the Gobar Gas Company (GGC). Other shareholders included Agricultural Development Bank of Nepal (ADB/N) and the Fuel Corporation of Nepal.

In 1985, UMN passed control of GGC to ADB/N, who had obtained funding from United Nations Development Programme (UNDP) and United Nations Capital Development Fund (UNCDF). They were able to offer subsidies to customers, so the demand for biogas grew. A mid-term evaluation provides an overview of the project (Fulford *et al.*, 1991). UNCDF requested technical support from Netherlands Development Organisation (SNV), which supplied two volunteers. A further evaluation of the UNDP-funded project was done by SNV (Leermaker, 1992). With the completion of the UNDP support, SNV took over the project, setting up the Biogas Support Programme (BSP) in 1992 (Karki *et al.*, 2007).

SNV has made available a large amount of literature on the biogas programme as it developed in Nepal. SNV started the Asia Biogas Programme (ABP) in 2003, to take a similar approach to Vietnam. The programme was taken to Bangladesh and Cambodia in 2006 and then Lao PDR in 2007 (Fulford *et al.*, 2012). The work of biogas extension has been developed further, as new programmes are being set up in many other countries across the world. Documents, such as project plans and evaluations, for many of these programmes are available from SNV.

However, the approach of SNV is only one of many ways in which biogas technology can be made available to large numbers of people. The total number of plants built in Nepal was over 0.25 million at the end of 2011, while the number claimed for India was 4.4 million and the number claimed for China was 42.8 million (Energy4All, 2012). The total number for the whole SNV biogas programme was 0.5 million by the end of 2012 (SNV, 2013).

The best biogas extension programmes in the world have been identified by Ashden.

This book continues the theme of its predecessor *Running a Biogas Programme*, but includes the various approaches that have been followed in different places. Lessons can be learned from the way these programmes are run, which can be applied in biogas extension programmes in other places.

Biogas in developing countries

The large scale programmes in countries such as China, India and Nepal were encouraged by the oil price jumps in the 1970s and 1980s, but they were originally inspired by a desire to find a replacement for wood as a domestic fuel for people in rural areas; this ensured these programmes continued expanding even after the oil price dropped. The need to find local and low-cost ways to generate energy in rural areas was seen as part of a wider set of development issues. Many development experts saw biogas and other renewable energy technologies offering part of the answer to development issues. Various organizations (UN, government and NGO) that were inspired by these development concerns, such as Intermediate Technology Development Group (ITDG) set up by Fritz Schumacher, later called Practical Action, became involved in extension work in renewable energy technologies including that of biogas. For example, Food and Agriculture Organisation of the United Nations (FAO) ran several courses on the subject (FAO, 1984).

In the 1980s, a growing concern was the overuse of fuel-wood for cooking which was leading to deforestation (Eckholm, 1975; Joseph and Hassrick, 1984). In mountainous areas, such as in Nepal and certain areas of China, deforestation leads to soil erosion and landslides, because tree roots bind together the soil on steep hillsides. The people who live in these areas often keep animals, such as cattle and pigs. They can use the dung from these animals to generate biogas for cooking, and replace their use of fuel-wood. The fertilizer value of the dung is enhanced, rather than lost, when it is passed through a biogas plant. Even in less mountainous areas, deforestation leads to loss of animal habitats and reduced biodiversity, as well as a reduction of soil fertility.

The use of biogas to replace fuel-wood for cooking was further encouraged by further concerns that smoke generated from burning wood causes ill health, as well as risks such as children falling into fires (Sefu, 1986; Smith, 1986). Biogas is a clean fuel that produces only carbon dioxide and water when it burns, so users do not suffer from the ill effects that they had when cooking on wood, such as irritation of the eyes, nasal passages and lungs from smoke pollutants, causing asthma and respiratory diseases. Another factor that reduced ill health was the addition of a latrine to the plant. In many rural areas, people do not have sanitation facilities, so have to use the fields. The use of a biogas plant allows the sewage wastes to be used as fertilizer, as previously, and the digestion process reduces the pathogens in the slurry (Poudel *et al.*, 2009). Medium temperature (mesophilic) digestion does not eliminate pathogens completely: some pathogens were reduced by 95 per cent, but others, such as viruses, by only 51 per cent (Marchaim, 1992). However, the pathogen reduction reduces the possibility of infection compared to the previous situation, where the sewage was not treated at all. If biogas effluent slurry is mixed with dry biomass and allowed to compost, higher temperatures can be reached, so the pathogens are further reduced.

In places where wood fuel is in short supply, people often dry animal dung for use as a fuel. The dung is usually mixed with straw and other agricultural residues, made into logs or flat plates and dried in the sun. The burning of dung means it cannot be used as fertilizer on the fields. The smoke generated from the burning of dung is even more smoky and noxious than smoke from the burning of wood or dry agricultural residues. The introduction of a biogas plant means that the same dung can be used both as a source of energy and a source of fertilizer. The effluent from a biogas plant can be mixed with other agricultural residues to form an even more effective compost (Ashden, 2007b, 2006a). However, biogas programmes in developing countries have placed less emphasis on the use of biogas effluent as fertilizer than it deserves.

In more recent years, a major motivation for the expansion of biogas programmes has been for the replacement of fossil fuels for cooking. The prices of cooking fuels, such as kerosene and LPG, have been subsidised in many countries to help poorer people pay for cooking. However, as the price of these fuels rises, the cost to government of the subsidies has increased. The Kyoto protocol and the influence of the United Nations Framework Convention on Climate Change (UNFCCC) have encouraged governments and other groups to look at biogas systems as a way to replace kerosene and LPG as cooking fuels.

As people in rapidly developing countries, such as India, become urbanized, their standards change and they become concerned about the cleanliness of their environment. The use of biogas systems as a way to effectively dispose of food waste and sewage is becoming a feature in several places, such as Mumbai (Shah, 2006), Pune (Ashden, 2006b), and the state of Kerala (Ashden, 2007a) in India. Such systems can be used at a domestic level, with individual families processing food wastes and/or sewage to generate energy for all or part of their own cooking needs. Biogas systems can also be used at an institutional level, in schools, hospitals, hostels (Idan, 2008), and prisons (Ashden, 2005b) and also at local authority level to process wastes from food markets and abattoirs (Ashden, 2007a).

Aspects of biogas technology in developed countries

Anaerobic digestion is used for several different applications in developed countries. These applications are seen as separate specialist applications, so are seen as separate subjects with their own literature. There is a lot of overlap between the different subjects and much can be learned from the literature in one subject area that can be applied in others. However, the subject areas have tended to develop somewhat in isolation from each other and there is some resistance from specialists in one area to consider ideas from another.

The main area of application for anaerobic digestion has been in sanitation; it is seen as one of several methods to process sewage. Modern drainage systems were built to combine wastewater from many sources, such as flush toilets (black water), washing facilities, such as baths and showers as well as from the washing of clothes and cooking and eating utensils (grey water), as well as rain water from drains. The result of mixing these different wastewaters together results in a very dilute slurry.

Sewage treatment plants are therefore designed for such dilute slurries. One approach is to use settling beds to separate the solid matter (sludge) from the rest of the water. The sludge is then often anaerobically digested, while the separated water is processed aerobically. Other approaches process the whole mix aerobically or use anaerobic systems. For example, the Upflow Anaerobic Sludge Blanket (Schellingkout and Collazos, 1991) concentrates and processes the waste simultaneously (Parawira, 2004). This is an area that has its own specialist literature.

The processing of industrial wastewaters is another area that is seen as distinct from sewage treatment, but has many close parallels. Wastewater from processes such as paper making, brewing and food processing often share similar characteristics to sewage, as they are often dilute, but have a high potential for pollution. Again this area has its own specialist literature (Stronach *et al.*, 1986).

Since the late 1980s, there has been concern about methane leaking from the anaerobic digestion that takes place in waste landfill sites. In many developed countries, domestic waste was all placed in a single bin and collected by local authorities. The waste was disposed of at landfill sites, holes that were naturally occurring or the result of removal of gravel or clay. The food waste in these holes decomposes and produces biogas. The gas can migrate into enclosed areas, where it can be ignited by a spark and cause an explosion (Deed, 2003; Gregson, 2000). Even if an explosive mixture does not form, the mixture of carbon dioxide and methane can replace the air in an enclosed space (such as an inspection pit or cellar in a building) and cause suffocation of people who enter the space (ATSDR, 2001). Since 1990s, all new landfill sites require the holes to be properly lined with an impervious membrane and the landfill gas collected and either flared off or burnt to generate energy (ATSDR, 2001). Older landfill sites must have vents added, so the gas can escape to the atmosphere, rather than migrate through the soil.

The use of landfill gas to fuel electric generators was developed in the UK in the late 1980s. Under the Non-Fossil Fuel Obligation (NFFO) scheme in the UK, 208 landfill gas projects were actually built to generate electricity, with a declared net capacity of 908 MW (REStats, 2012). Gas from landfill sites was also used in furnaces to generate heat for industrial operations, such as firing kilns for brick and tile making. Landfill gas is another specialist area that has generated its own literature. Under different government schemes, the generation was increased in the UK to 1,067 MW.

As the idea that wastes need to be recycled has developed since the mid-1980s, there has been a growth in the encouragement by local and national governments for the separation of wastes at source. The EU has placed a lot of pressure to increase recycling of wastes and people in many European countries, notably Germany, Holland, and Belgium, have led the way in the collection of metal, glass, plastics and food wastes from separate domestic bins. The need to process the food wastes has encouraged the development of both aerobic (composting) and anaerobic systems (Biosantech *et al*, 2013). In recent years, several in-vessel waste processing systems have been developed for municipal solid waste. DRANCO, Valorga, and Kompogas (Rapport *et al.*, 2008; Wellinger, 2008) are examples of in-vessel waste processing systems.

On-farm biogas systems in developed countries

The prime aim of this book is to look at the way a biogas programme can be run in a less developed country. However, some reference should be made to the recent development of the use of anaerobic digestion in Europe and the USA to process waste and to generate energy. These programmes have been well documented, especially by the International Energy Authority (IEA) 'Task 37: Energy from Biogas and Landfill Gas' (Wellinger and Murphy, 2013).

During the oil price-hikes of the 1970s and 1980s there was enthusiasm for building farm-scale biogas plants in various countries. For example, in the UK, between 45 and 50 plants were built. By 1993, 25 were still working (Baldwin, 1993). In Germany, 73 farm-scale biogas plants were built by 1982 (Pellmeyer, 2009). The drop in oil prices in the mid-1980s meant that many companies building these plants either went out of business or turned to other products. The plant owners used their plants for a time, but many gave up when the suppliers were unable to offer repair and maintenance services.

However, different German aid agencies, such as BORDA, GATE and GTZ were able to maintain their interest in the technology through offering their technical services to projects in developing countries (Sasse, 1988; Sasse *et al.*, 1991). This involvement meant that Germany was able to keep a keen interest in biogas technology and was able to develop this technology quickly, as the demand for it has grown again (GTZ, 2010).

In the growing concern for the environment, Germany has led the way in developing larger-sized biogas units, after 2000 when the government introduced attractive grid feed-in tariffs for electricity generated from on-farm systems (Kram, 2007). The rules were expanded in 2004 to include biogas from crops. By the end of 2011, there were 7,320 agricultural biogas systems in Germany (Linke, 2013) with a total electricity generation capacity of 2,997 ME and the total at the end of 2013 was anticipated to be 7,874 plants with a capacity of 3,364 MW. Agricultural biogas generated 15 per cent of the total electricity from renewable energy in 2012. Most of these plants are fed with a mixture of manure plus an energy crop, usually maize fodder silage (Braun and Weiland, 2010). Material that has not been passed through a cow gives a much higher gas production per kilogram of input material.

The use of on-farm biogas systems has expanded in parts of the rest of Europe, as well as in Germany. There is also growth in the use of this technology in the USA, Canada, Australia, and New Zealand. Various groups have sprung up to share information and act as trade organisations for the work of installing biogas systems. IEA Task 37 acts to coordinate the work of many national groups.

In the UK, the Renewable Energy Association has a Biogas Group that acts as a trade association and pressure group. The UK is beginning to develop a programme for biogas and by the end of 2010 had about 40 farm-scale digesters, three designed to process municipal food wastes and 18 for wastes from food-processing industries. While some of these plants were left over from the 1980s programme and several more were built using German technology, a few companies, such as BiogenGreenfinch, use technology which they have developed within the UK (Harwood, 2010).

In the USA, the foundation of on-farm anaerobic digestion research was done at Cornell University, where a group led by W.J. Jewell built a prototype plant in the late 1970s (Jewell *et al.*, 1978). About 100 on-farm biogas plants were built, although only 10 survived until 1995 (Roos, 2009). Many of these plants consisted of rubberized covers placed over existing slurry pits to collect the gas. Beef cattle in the USA are usually allowed to range free, but herded into large feedlots to fatten before they are slaughtered. These feedlots generate large volumes of dung, which is collected into slurry pits, so the use of a flexible cover provided a quick and easy way to collect the biogas generated. These rubberized covers had a fairly short lifetime (a few years) and the drop in energy costs in the mid-1980s meant that farmers had little incentive to replace the covers when they failed. The ones that survived were mainly on dairy or pig farms.

The United States' Environmental Protection Agency (EPA), together with the US Department of Agriculture, and the US Department of Energy, set up the AgStar project in 1994 to provide US farmers with a centralized resource for detailed information on biogas systems. AgStar produced a handbook on biogas systems (Roos *et al.*, 2004), which included software to help farmers assess how the technology can benefit their operations. The number of biogas systems in the USA was 176 at the end of 2011, generating 541 GWh of energy (AgStar, 2010), mainly as electricity, but some is used to generate heat.

Biogas extension work

This book aims to act as a primer on the use of anaerobic technology to provide a source of energy, mainly in developing countries. It uses the experience of various biogas extension projects to show ways in which this technology could be used more widely.

References

AgStar (2014) US Anaerobic Digestion Status Report, US EPA <www.epa.gov/agstar/projects/index.html> [accessed 17 July 2014].

Alleman, J.E. (1982) *The History of Fixed-Film Wastewater Treatment Systems*, West Lafayette, Indiana: Purdue University. <http://web.deu.edu.tr/atiksu/ana52/biofilm4.pdf> [accessed 17 July 2014].

Ashden (2006a) 'Appropriate Rural Technology Institute: Compact Digester for Producing Biogas from Food Waste' [Website - case study] Ashden <www.ashden.org/winners/arti06> [accessed 17 July 2014].

Ashden (2005a) 'Biogas Sector Partnership, Nepal' [Website - case study] Ashden <www.ashden.org/winners/BSP> [accessed: 17 July 2014].

Ashden (2007a) 'Biotech: Management of Domestic and Municipal Waste at Source Produces Biogas for Cooking and Electricity Generation' [Website - case study] Ashden <www.ashden.org/winners/biotech> [accessed 17 July 2014].

Ashden (2005b) 'Kigali Institute of Science, Technology and Management (KIST), Rwanda' [Website - case study] Ashden <www.ashden.org/winners/kist05> [accessed 17 July 2014].

Ashden (2010a) 'MARD/SNV, Vietnam; Biogas for Smallholders in Vietnam' [Website - case study] Ashden <www.ashden.org/winners/MARD10> [accessed 17 July 2014].

Ashden (2006b) 'Shaanxi Mothers, China; Domestic Biogas for Cooking and Lighting' [Website - case study] Ashden <www.ashden.org/winners/shaanxi> [accessed 17 July 2014].

Ashden (2012) 'SKDRDP, India: Enabling the Poor to Make Informed Energy Choices' [Website - case study] Ashden <www.ashden.org/winners/skdrdp12> [accessed 17 July 2014].

Ashden (2007b) 'SKG Sangha, India; Biogas for Cooking Plus Fertiliser from Slurry' [Website - case study] Ashden <www.ashden.org/winners/skgsangha> [accessed 17 July 2014].

Ashden (2010b) 'Sky Link Innovators, Kenya' [Website - case study] Ashden <www.ashden.org/winners/Skylink10> [accessed 17 July 2014].

Ashden (2006c) 'VK Nardep, India: Adding Value to the Residue from Biogas Plants' [Website - case study] Ashden <www.ashden.org/winners/vknardep> [accessed 17 July 2014].

Ashden (2006d) 'VK-NARDEP, India: Multiple benefits from biogas' [Web site - case study] Ashden <www.ashden.org/winners/vknardep> [accessed: 17/07/2014].

ATSDR (2001) *Landfill Gas Primer – Safety & Health an Overview for Environmental Health Professionals*, for Toxic Substances and Disease Registry, USA: Department of Health and Human Services Agency Assessment and Consultation. <www.atsdr.cdc.gov/HAC/landfill/html/intro.html> [accessed 17 July 2014].

Baldwin, D.J. (1993) 'Anaerobic digestion in the UK: a review of current practice', in *Energy from Biomass: Anaerobic Digestion for Biogas. Summaries of the Biomass Projects Carried Out as Part of the Department of Trade and Industry's New and Renewable Energy Programme*, Silsoe: ADAS - ETSU. <https://wedc-knowledge.lboro.ac.uk/details.html?id=6374> [accessed 17 July 2014].

Biosantech, T.A.S.; Rutz, D.; Janssen, R. and Drosg, B. (2013) '2 - Biomass resources for biogas production'. In A. Wellinger, J. Murphy, & D. Baxter, eds. *The Biogas Handbook.* , Woodhead Publishing, pp. 19–51. Cambridge UK <http://dx.doi.org/10.1533/9780857097415.1.19> [accessed 30 July 2014].

Braun, R. and Weiland, P. (2010) 'Biogas from Energy Crop Digestion, IEA Bioenergy Task 37' <www.iea-biogas.net/files/daten-redaktion/download/energycrop_def_Low_Res.pdf> [accessed 17 July 2014].

Brundtland, G.H. (1987) 'Our Common Future, Report of the World Commission on Environment and Development, World Commission on Environment and Development', Published as Annex to General Assembly document A/42/427, *Development and International Co-operation: Environment* <http://conspect.nl/pdf/Our_Common_Future-Brundtland_Report_1987.pdf> [accessed 17 July 2014].

Bulmer, A., Finlay, J., Fulford, D.J. and Lau-Wong, M.M. (1985) *Biogas: Challenges and Experience from Nepal* Vols. I and II, Butwal, Nepal: United Mission to Nepal <www.kingdombio.com/Biogas-vol-I.pdf> [accessed 17 July 2014].

Burian, S.J., Nix, S.J., Pitt, R.E. and Durrans, S.R. (2000) 'Urban Wastewater Management in the United States: Past, Present, and Future', *Journal of Urban Technology* 7: 33–62 <http://dx.doi.org/10.1080/713684134>. [accessed 17 July 2014].

Buswell, A.M. (1936) 'Anaerobic Fermentations, State of Illinois, State Water Survey, Bulletin No 32' <www.isws.illinois.edu/pubdoc/B/ISWSB-32.pdf> [accessed 17 July 2014].

Buswell, A.M. (1930) *Production of Fuel Gas by Anaerobic Fermentations, State of Illinois, State Water Survey, Circular No 10* www.isws.illinois.edu/pubdoc/C/ISWSC-10.pdf [accessed 17 July 2014].

Caldwell, A. (2008) 'Sheffield's Sewer Gas Destructor Lamps' [Website] <http://alancordwell.co.uk/Legacy/misc/webb.html> [accessed 17 July 2014].

Chen, Y., Yang, G., Sweeney, S. and Feng, Y. (2010) 'Household biogas use in rural China: A study of opportunities and constraints', *Renewable and Sustainable Energy Reviews* 14: 545–549 <http://dx.doi.org/10.1016/j.rser.2009.07.019>.

Cooper, P.F. (2002) 'Historical aspects of wastewater treatment'. in *Decentralised Sanitation and Reuse: Concepts, Systems and Implementation*, IWA Publishing, London.

Deed, C. (2003) 'Guidance on the management of landfill gas', UK Environment Agency, Rotherham, UK <www.sepa.org.uk/waste/waste_regulation/idoc.ashx?docid=e13df631-957f-4b0b-8487-4f2df89c8421&version=-1> [accessed 17 July 2014].

Eckholm, E. (1975) *Other Energy Crisis*, Worldwatch Institute, Washington DC.

Energy4All (2012) 'Brief progress and planning report the Working Group on Domestic Biogas' <www.hedon.info/tiki-download_item_attachment.php?attId=443> [accessed 17 July 2014].

Enongene, G.N. (2003) The Enzymology of Enhanced Hydrolysis within the Biosulphidogenic Recycling Sludge Bed Reactor, PhD Thesis, Rhodes University. http://eprints.ru.ac.za/54/ [accessed 17 July 2014].

FAO (1984) 'Biogas: what it is; how it is made; how to use it' <www.pssurvival.com/ps/biogas/Better_Farming_Series_32_Biogas_2_Building_A_Better_Biog as_Unit_Fao_1986.pdf> [accessed 17 July 2014].

Frerichs, R.R. (2001) 'History, maps and the internet: UCLA's John Snow site' *SoC BULLETIN* 34<www.ph.ucla.edu/epi/snow/socbulletin34%282%293_7_2001.pdf> [accessed 17 July 2014].

Fulford, D. (1988) *Running a Biogas Programme: A Handbook*, London: Practical Action Publications (Intermediate Technology Publications). <http://developmentbookshop.com/runninga-biogas-programme-pb> [accessed 22 July 2014].

Fulford, D., Devkota, G.P. and Afful, K. (2012) *Evaluation of Capacity Building in Nepal and Asia Biogas Programme,* Hanoi, Vietnam: Kingdom Bioenergy Ltd for SNV, Reading UK. <www.kingdombio.com/Final%20Report%20-%20whole.pdf> [accessed 17 July 2014].

Fulford, D., Poudal, T.R. and Roque, J. (1991) *Evaluation of On-going Project: Financing and Construction of Biogas Plants*, United Nations Capital Development Fund, New York.

Gikas, P., Georgakopoulos, A. and Droumbogianni, I. (2004) *Commissioning Strategy for the Anaerobic Sludge Digesters at the Athens Wastewater Treatment Plant in Psyttalia*, Athens: Ministry of Environmental Planning and Public Works. <www.srcosmos.gr/srcosmos/showpub.aspx?aa=8313> [accessed: 17/07/2014].

GTZ (2010) *Guide to Biogas: from Production to Use,* 5th edn, Germany: Fachagentur Nachwachsende Rohstoffe e. V. (FNR) [accessed 17 July 2014].

Harris, P. (2008) 'Beginners Guide to Biogas' [Web page - guide] <www.adelaide.edu.au/biogas/> [accessed 17 July 2014].

Hartley, H. (1960) 'The Wilkins Lecture. Sir Humphry Davy, Bt., P.R.S. 1778–1829' *Proceedings of the Royal Society of London. Series A, Mathematical and Physical Sciences* 255: 154–180.

Harwood, O. (2010) UK Country Report [Presentation] Den Bosch, Netherlands <www.iea-biogas.net/countryreports.html?file=files/daten-redaktion/download/publications/countryreports/2010/UK_Country_Report_11-2010.pdf> [accessed: 17 July 2014].

Hobson, P.N. and Bousfield, S. (1981) *Methane Production from Agricultural and Domestic Wastes*, Applied Science Publishers, London.

Hopkins, P. (2007) *Oil and Gas Pipelines: Yesterday and Today*, International Petroleum Technology Institute, Penspen Ltd, ASME, Houston, Texas <www.penspen.com/Downloads/Papers/Documents/OilandGasPipelines.pdf> [accessed 17 July 2014].

Idan, J.A. (2008) 'Integrated Sewage and Solid Organic Waste-to-Energy Project' [Website – case study] Accra, Ghana <www.biogasonline.com/downloads/integrated_waste_to_energy_project.pdf> [accessed 17 July 2014].

Jewell, W.J., Capener, H.R. and Dell'orto, S. (1978) *Anaerobic Fermentation of Agricultural Residue: Potential for Improvement and Implementation. Final report*, Cornell University for US, Ithaca, USA DOE [accessed 17 July 2014].

Joseph, S. and Hassrick, P. (1984) *Burning Issues, Implementing Pilot Stove Programmes, A Guide for East Africa*, UNICEF, IT Publications Rugby UK.

Kansal, A., Rajeshwari, K.V., Malini Balakrishnan, Lata, K. and Kishore, V.V.N. (2004) 'Anaerobic digestion technologies for energy recovery from industrial wastewater - a study in Indian context', *TERI Information Monitor on Environmental Science* 3: 67–75, New Delhi.

Karki, A.B., Shakya, I., Dawadi, K.D. and Sharma, I. Eds (2007) *Biogas Sector in Nepal; Highlighting Historical Heights and Present Status*, Kathmandu: Nepal Biogas Promotion Group (NBPG).

Kossmann, W., Habermehl, S., Hörz, T., Krämer, P., Klingler, B and Klopotek, F. v. (1999) *Biogas Digest Volume I: Biogas Basics*, Wiesbaden, Germany: ISAT and GTZ <www.susana.org/langen/library?view=ccbktypeitem&type=2&id=526> [accessed 17 July 2014].

Kram, J.W. (2007) 'A New Day for Biogas: Germany Leads the Way in Europe', *Biomass Magazine* 1: 24-29 <http://issuu.com/bbiinternational/docs/bmmjune.07_print?e=2317201/7370335> [accessed 17 July 2014].

Kurian, P.K. (2004) 'Socio-Economic and Environmental Impact of Biogas Programme.' Kerala Research Programme On Local Level Development, Centre For Development Studies, Thiruvananthapuram. <www.cds.ac.in/krpcds/report/pkkurian.pdf > [accessed 17 July 2014].

Leermaker, M. (1992) *Extension of Biogas in Nepal - Parts 1, 2 and 3*, SNV, The Hague, Netherlands <www.snvworld.org/en/download/publications/extension_of_biogas_in_nepal_theory_and_practice_1992_0.pdf> [accessed 22 July 2014].

Linke, B. (2013) 'Country Report, Germany' [Presentation] <www.iea-biogas.net/countryreports.html?file=files/daten-redaktion/download/publications/countryreports/may2013/Country_Report_Germany_April_2013.pdf> [accessed 17 July 2014].

Marchaim, U. (1992) *Biogas Processes for Sustainable Development*, Kiryat Shmona, Israel: FAO., Rome <www.fao.org/docrep/t0541e/T0541E00.HTM> [accessed: 17 July 2014].

Meadows, D.H., Meadows, D.L., Randers, J. and Behrens, W.W. III (1972) *Limits to Growth*, Club of Rome.

Parawira, W. (2004) 'Anaerobic Treatment of Agricultural Residues and Wastewater Application of High-Rate Reactors', Sweden: PhD, Lund University. <http://lup.lub.lu.se/luur/download?func=downloadFile&recordOId=467675&fileOId=1472236> [accessed: 17 July 2014].

Parkin, G. (1986) 'Fundamentals of Anaerobic Digestion of Wastewater Sludges.' *J. Environ. Eng* 112: 867–920 <http://dx.doi.org/10.1061/(ASCE)0733-9372(1986)112:5(867)> [accessed 22 July 2014].

Pellmeyer, J. (2009) 'Biogas – a German Success Story', *International Sustainable Energy Review* 4 <www.internationalsustainableenergy.com/319/iser-magazine/past-issues/biogas-a-germansuccess-story/> [accessed: 17 July 2014].

Poudel, R.C., Joshi, D.R., Dhakal, N.R. and Karki, A.B. (2009) 'Evaluation of Hygienic Treatment of Biowastes by Anaerobic Digestion in Biogas Plants', *Nepal Journal of Science and Technology* 10: 183–188 <http://dx.doi.org/10.3126/njst.v10i0.2958>.

Rapport, J., Zhang, R., Jenkins, B.M. and Williams, R. (2008) *Current Anaerobic Digestion Technologies Used for Treatment of Municipal Organic Solid Waste*, Sacramento: California Integrated Waste Management Board. <www.calrecycle.ca.gov/publications/Documents/1275%5C2008011.pdf> [accessed 17 July 2014].

REStats (2012) 'Chapter 6: Renewable sources of energy', in *Digest of United Kingdom Energy*, London *Statistics* (DUKES), Department of Energy and Climate Change, UK. <http://webarchive.nationalarchives.gov.uk/20130109092117/http://decc.gov.uk/assets/decc/11/st ats/publications/dukes/5956-dukes-2012-chapter-6-renewable.pdf > [accessed 17 July2014].

Roos, K. (2009) *Expansion of the U.S. Digester Market in the Dairy and Pork Sector*, US Environment Protection Agency (EPA), AgStar, Washington DC <www.epa.gov/agstar/documents/conf09/kurt_agstar_market_expansion_presentationfinal%20.pdf> [accessed: 17 July 2014].

Roos, K.F., Martin, J.B.J. and Moser, M.A. (2004) *AgSTAR Handbook: A Manual for Developing Biogas Systems at Commercial Farms in the United States*, 2nd edn, US EPA and ERG Inc , AgStar, Washington DC <www.epa.gov/agstar/documents/AgSTAR-handbook.pdf> [accessed 17/07/2014].

Sasse, L. (1988) *Biogas Plants*, GATE, GTZ Wiesbaden, Germany <www.gateinternational.org/documents/publications/webdocs/pdfs/g34bie.pdf> [accessed 17 July 2014].

Sasse, L., Kellner, C. and Kimaro, A. (1991) *Improved Biogas Unit for Developing Countries*, Eschborn: GATE, GTZ Wiesbaden, Germany <http://www2.gtz.de/Dokumente/oe44/ecosan/en-improvedbiogas-unit-1991.pdf> [accessed 17 July 2014].

Schellingkout, A. and Collazos, C.J. (1991) 'Full-Scale Application of the UASB Technology for Sewage Treatment', *Water Science and Technology* 25: 159–166 <www.iwaponline.com/wst/02507/wst025070159.htm> [accessed 17 July 2014].

Schumacher, E.F. (1973) *Small is Beautiful: Economics as if People Mattered*, Hartley & Marks, Vancouver, Canada.

Sefu, A. (1986) 'Improved stoves: Safety is important too', *Boiling Point* 13 <www.hedon.info/BP13_Improved Stoves:SafetyIsImportantToo?bl=y> [accessed 17 July 2014].

Shah, O.P. (2006) Sustainable Waste Processing in Mumbai: Using the Nisargruna Technology. MSc Thesis. Borås, Sweden: Hogskolan i Borås. <http://www.kingdombio.co.uk/ShahThesis.pdf> [accessed 17 July 2014].

Smith, K.R. (1986) 'Cookstove Smoke and Health', *Boiling Point* 13 <www. hedon.info/BP13_CookstoveSmokeAndHealth?bl=y> [accessed 17 July 2014].

SNV (2013) 'Milestone 500,000 biodigesters reached' SNV World [Website - magazine] The Hague, Netherlands <www.snvworld.org/en/regions/africa/ news/milestone-500000-biodigesters-reached> [accessed 17 July 2014].

Stronach, S.M. Rudd, T. and Lester, J.N. (1986) *Anaerobic Digestion Processes in Industrial Wastewater Treatment, Biotechnology Monographs.* Springer, New York <http://link.springer.com/book/10.1007%2F978-3-642-71215-9> [accessed 17 July 2014].

UN (1992) *The United Nations Framework Convention on Climate Change*, FCCC/ INFORMAL/84 GE.05-62220 (E) 200705 United Nations, Bonn, Germany. <http://unfccc.int/resource/docs/convkp/conveng.pdf> [accessed 17 July 2014].

UN (2008) *Fact sheet: The Kyoto Protocol*, UNFCCC, Bonn, Germany https:// unfccc.int/files/press/backgrounders/application/pdf/fact_sheet_the_ kyoto_protocol.pdf [accessed 17 July 2014].

UKEA (2014) *LFTGN03: guidance on the management of landfill gas*, UK: Environment Agency, Rotherham, UK http://www.sepa.org.uk/waste/waste_regulation/ idoc.ashx?docid=b0b554c4-3ed3-49d0-85c5-bde23b82de60&version=-1 [accessed 17 July 2014].

Wellinger, A. (2008) 'IEA Task 37' [Website - presentation] IEA Task 37 meeting, Ludlow, UK <www.ieabiogas. net/files/daten-redaktion/download/ publications/workshop/4/presentation_wellinger4-08.pdf> [accessed 17 July 2014].

Wellinger, A. and Murphy, J. (2013) *Biogas Handbook: Science, Production and Application,* D Woodhead Publishing Series in Energy, Cambridge, UK <www.iea-biogas.net/biogas-handbook.html> [accessed 17 July 2014].

Wolfe, R.S. (2004) 'Pistola di Volta: Recreating Volta's dramatic 19th-century displays of energy released from methane provides insights into anaerobic microbial metabolism.' *ASM News* 70: 18 <http://forms.asm.org/microbe/ index.asp?bid=24030> [accessed 17 July 2014].

CHAPTER 2
Biogas history in developing countries

Abstract

Three large biogas programmes in the developing world, in China, India, and Nepal, have been growing since the 1970s. Some of the smaller biogas programmes in other countries also started in the 1970s but have not grown at the same rate. The history of the three large programmes offers insights into the reasons for their success. Programmes in countries in the rest of Asia, in Africa and in Latin America are considered so that comparisons can be made. Biogas programmes in many of these countries are now being influenced by lessons learned from the larger programmes.

Keywords: biogas, anaerobic digestion, history

Biogas programme in China

The Chinese programme started in 1920 when an entrepreneur, Luo Guorui, invented and built a 8 m³ biogas plant and sold units under the name of Chinese Guorui Biogas Company in the area around Shanghai (Kangmin, 2006). An emphasis on the value of biogas plant effluent as fertilizer and the need to find a replacement for wood fuel to reduce deforestation inspired a programme that started in Wuchang in 1958 and spread across the country. There were reports of 7 million plants built under this programme by 1978 (Karki, 1996). The work had a strong push between 1975 and 1979 from Chairman Mao in a national programme entitled: 'Biogas for Every Household'. This programme generated a keen interest in biogas technology, especially with the publication of a translation of *A Chinese Biogas Manual* by IT Publications (van Buren, 1979).

Unfortunately, this programme was based on political activists encouraging villagers to build do-it-yourself digestion systems, so the plants were not very reliable. The emphasis was on quantity, not quality (Harter, 2010). The Cultural Revolution created further difficulties because many people with technical expertise were forced to leave universities and research institutions to work in the fields. By 1978 only 3 million of the 7 million plants installed were still in operation (Karki, 1996). It is possible, though, that technically trained people who were forced to work in local communities were able to suggest the improvements in the designs that came out of this programme.

As attitudes changed, the need for a nationally coordinated approach was recognized (Wu, 2003). The Chengdu Biogas Scientific Research Institute of the Ministry of Agriculture (Liu and Zhao, 2008) was established in 1979, along with the Asia-Pacific Regional Biogas Research and Training Centre as a means

http://dx.doi.org/10.3362/9781780448497.002

to train biogas technicians in making good-quality biogas systems. A series of standards were drawn up for the building of biogas plants and other equipment (Chen *et al.*, 2012; NSPRC, 1985) to ensure that the programme would continue on a firmer base. The Chinese government introduced a cash subsidy for biogas plants to cover the extra cost of meeting high construction standards.

The more centralized programme continued, following a dip in numbers in the early 1980s. People were slow to regain confidence in the technology, so plants were installed at a much slower, but steadier, rate. The number of plants in China increased very slowly reaching a total of 8 million in 2000 (Tu, 2010). Although many plants were being built, they were mainly replacing plants that had failed. The average life-span of a biogas plant in China was only about 4.5 years (Tu, 2010).

Since the year 2000, the number of biogas plants in China has increased as a result of a combination of national government initiatives and local advocates. The government's ninth and tenth Five-Year Plans (2000 to 2010) included increasing investment in biogas (Chen *et al.*, 2012). A local example of increasing investment is a group of women, Shanxi Mothers, who encouraged the use of biogas in one locality (Ashden, 2006a). From 2003, local and national initiatives resulted in a rapid increase in the building of biogas plants (Zuzhang, 2013), resulting in some 40 million plants built by the end of 2011. A more accurate figure from the Energy4All programme of ADB (Energy4All, 2012) is 43.8 million plants by the end of 2011.

The Chinese approach emphasises integration between different components (Chen *et al.*, 2010), so many of these systems are of the three-in-one format (pigsty, latrine, and biogas plant) that work together to generate energy and compost. Some are four-in-one (pigsty, latrine, digester, and greenhouse), especially in colder areas (CNSS, 2011), where the greenhouse helps to keep the temperature of the digester above ambient, as well as allowing plants to grow better in cold weather. There are also five-in-one systems, which include fishponds into which the slurry is fed to encourage the growth of algae as fish food.

A recent innovation is that of prefabricated biogas plants made of plastics (Cheng *et al.*, 2013). These include the use of flexible-bag digesters, but also underground dome digesters made from fibre reinforced plastics (FRP) and welded rigid plastics (HDPE, PVC, ABS, etc.). These systems are now being exported to places such as Vietnam and Bangladesh.

There is also a new interest in China in the digestion of wastes from food processing and larger agricultural operations, such as dairy units and animal fattening units (Jiang *et al.*, 2011). These larger digesters are made from steel, plastic or concrete cast in metal formwork (Wilson, 2013), with some projects built in cooperation with German companies.

Biogas programme in India

Following the successful biogas systems in Bombay, there was ongoing research work in this area in the 1920s in places such as the Indian Institute of Technology in Bangalore. Attempts were made in India in the 1930s to design a technology that could generate biogas from animal manure, but the systems that were made were much too expensive for use by Indian farmers (Lawbuary, 2000). In 1951, J.J. Patel in Bombay developed the Gramalaxmi design of biogas plant (NIIR, 2004) which was chosen by the Khadi and Villages Industries Commission (KVIC) to launch a project to encourage the use of biogas in Indian villages in the 1960s. In 1961, the Planning Research and Action Division (PRAD), set up by the Uttar Pradesh State government (Breslin *et al.*, 1980), set up the Gobar Gas Research Centre in Ajitmal in Uttah Pradesh. Ram Bux Singh (Singh, 1973) publicized the idea of biogas across India and the rest of the world.

Following the oil price-hikes of the 1970s, several other organizations, both government and NGOs, became involved in biogas initiatives. KVIC acted as the lead organization, offering training courses and running extension programmes. PRAD developed a design of biogas plant based on the designs that were emerging from China in the early 1980s and called it the Janata (People's) design (Chauhan and Srivastava, 2012). An influential Indian NGO, Action For Food Production (AFPRO) also adapted ideas from the Chinese programme and came up with a design that did not require formwork for its construction. They called it the Deenbandhu (farmer's friend) design. Both PRAD and AFPRO had effective training programmes for technicians on how to make both designs of fixed-dome plant. The Deenbandhu plant was seen as more appropriate in situations where bricks were easily available, but the Janata design was more suitable for building from concrete.

In 1981 the government of India launched The National Project on Biogas Development (NPBD) which became part of Department of Non-conventional Energy Sources (DNES) when it was created in 1982 (Kurian, 2004). The Department became a Ministry ten years later (MNES). The National Dairy Development Board (NDDB) was also very active in promoting biogas plants. One of the main actions of the NPBD was to establish subsidies for biogas plants. The Ministry (which has now changed its name to Ministry of New and Renewable Energy – MNRE) makes agreements with the various extension agencies as to the number of plants they will install each year. In the early 2000s, the main agencies installing biogas plants were state government departments (of which PRAD is one) and energy development corporations. A study of NPBD by Programme Evaluation Organization of the Planning Commission of India (Pal, 2002) suggested that KVIC is still involved, although at a much lower level (7 per cent of the sample), with NGOs taking a minor role. Since the early 2000s, NGOs and small commercial companies have taken a larger role, although the quality of the plants built is still very variable.

Under the NPBD scheme, MNRE pays subsidies to the agencies, which then reduces the price they charge for each plant, by the amount defined by NPBD, to the farmers for whom they are installing them. State governments often add an extra subsidy on top of the national government one. The Indian banks, especially the Grameen (rural) banks, offer loans to farmers to enable them to pay their part of the cost of a plant. The government also offers subsidy on the interest on these loans.

At the end of 2009, MNRE claimed that a total of 4.18 million biogas plants had been set up in India under the National Biogas and Manure Management Programme (NBMMP), out of a government defined figure of 12 million for the potential construction of cattle-dung-based rural plants. NBMMP also trains masons and other biogas staff in nine main centres across India. The success rate of installing and maintaining biogas plants in India seems variable. Some agencies have been very successful and can claim that close to 100 per cent of their plants are still being used by farmers (Bhat and Chanakya, 2001). Other groups have been much less successful, either because their masons have not been properly trained or the agency has tried to cut corners. A study in 2002 of 615 plants suggested about 55 per cent of biogas plants in India continued to work effectively (Pal, 2002). Some of the more successful programmes are continuing to work to rehabilitate failed plants built by the less successful programmes (Vasudeo, 1993).

SKG Sangha (Ashden, 2007a) is notable as being an NGO with a strong emphasis on high-quality control and follow-up procedures. By the end of 2013, SKG Sangha had built over 125,000 biogas plants in South India. They also run extension programmes in Egypt and Mali, and have built pilot plants in places such as Kenya, Ghana, Honduras, Nicaragua, and Madagascar. SKG Sangha has been very successful in getting carbon-offset finance from both the regulated and voluntary carbon markets.

There is a new interest in processing food wastes in India and groups are looking at different scales of operation, from domestic to large scale. ARTI (Ashden, 2006b) and Biotech Ltd in Kerala (Ashden, 2007b; Biotech, 2008) are selling small plants to urban and suburban families to convert domestic food waste into cooking gas. Biotech Ltd also makes plants for institutions, such as schools and hostels, which digest both food wastes from the kitchens and sewage from the toilets. The third area is the use of market and other municipal wastes to make gas that can be used to generate electricity. By the end of 2009, Biotech Ltd had built 15 waste-to-electricity plants for markets and local authorities.

Bhabha Atomic Research Centre (BARC) in Mumbai, India is another organization looking at municipal solid waste (MSW) and it developed the Nisargruna plant (Shah, 2006). This is a two-stage digester that processes food waste from institutions and local authorities. The food waste is chopped finely in a crushing machine before it is hydrolized in a thermophyllic first stage digester which is solar heated. The pre-digested liquor then flows into a conventional floating drum second stage digester in which biogas is generated. Dr Kale of BARC has been in negotiation with various groups to commercialize this technology.

Biogas programme in Nepal

The pioneer of biogas in Nepal was Father B.R. Saubolle (Saubolle, 1983), who built a demonstration plant in 1955. In 1968, an exhibition at Kathmandu exhibited a biogas plant built by KVIC from India (Devkota, 2007). The Department of Agriculture declared 1974/75 to be an agricultural year and set a programme with the Agriculture Development Bank of Nepal (ADB/N) to install 250 biogas plants to the KVIC floating drum design with interest-free loans. Development and Consulting Services (DCS) working with Butwal Engineering works (BEW) became a subcontractor to this programme and built 95 plants. The involvement of DCS encouraged ADB/N and the Fuel Corporation of Nepal to set up Gobar Gas and Agriculture Equipment Development Company P. Ltd (GGC) with the United Mission to Nepal (UMN), the parent organisation of DCS, in 1977. DCS continued to offer research back-up to the company, with finance from USAID (Fulford, 1988). A team of expatriate experts developed new designs of biogas plant and associated equipment that could be used in the locally managed commercial extension programme.

The programme started slowly, with only a few hundred plants built each year. The research programme made improvements to the floating drum design, but it was discovered that the steel drums were rusting badly. Various other designs were tested, such as a floating ferro-cement gasholder, several fixed-dome designs and a tunnel design. In 1980, a modification of the Chinese fixed-dome design was developed, which has since been adapted to become the main plant that was built by GGC (GGC 2047 design). The research programme also developed a gas stove that was manufactured in Nepal.

ADB/N had a strong interest in using biogas to drive engines for irrigation and agro-processing as a way to help local farmers reduce the use of expensive, imported fossil fuels. Slow running, small (5 kW) dual-fuel engines, such as those that were available from Kirloskar at the time, ran very well on biogas, using 20 per cent diesel to provide ignition. A gas carburettor was developed that allowed faster running small diesels also to run as dual-fuel engines (Fulford, 1988). A larger design of plant (34 m^3 working volume) was required to supply gas to power dual-fuel engines. By 1986 GGC had installed 60 of these larger plants with engines for irrigation and agro-processing. The problem with such systems was that, in many cases, several families had to supply dung for the larger plant, which was run on a cooperative basis. Cooperation between the families proved difficult, so many of these plants failed for social reasons (Devkota, 2001).

The introduction of the fixed-dome design, which was about 30 per cent cheaper than the floating drum design, allowed the biogas plant market to grow slightly more quickly. In the late 1980s, UMN withdrew from the biogas project and ADB/N had funding from UNDP (United Nations Development Fund) and UNCDF (United Nations Capital Development Fund) to continue installing biogas systems. The UNCDF grant had provision for subsidies to be introduced for biogas plants. A 25 per cent subsidy suddenly increased the demand from farmers for biogas plants. ADB/N discovered they were exceeding their targets for building biogas plants each year (Fulford *et al.*, 1991). The total number of plants built in Nepal by 1992 was 12,000, mainly of the GGC fixed-dome design.

An advisor in UNCDF realized that GGC, as a single company, was not able to manage the increasing demand effectively, and recommended that SNV (Stichting Nederlandse Vrijwilligers) Netherlands Development Organisation, a Netherlands aid group, become involved. SNV set up the Biogas Support Programme (BSP), which licensed contractors to install plants (Bajgain, 2006). SNV also took over the financing of the biogas subsidy programme with help from the Netherlands and German governments. Staff of the Gobar Gas Company, who had already gained good experience, led the formation of 52 biogas installation companies. Other building contractors who expressed an interest in being licensed to install biogas plants were thoroughly trained by BSP. A selection of plants built by each installer was inspected by experienced BSP staff to check the quality of construction (Devkota, 2007). The installer must rebuild poor-quality plants or face losing their license. Plant subsidies can only be provided for plants built by licensed installers, therefore contractors had a strong incentive to maintain high quality. Regular follow-up visits by BSP have provided records suggesting that 98 per cent of plants built under the scheme were still working five years after they have been built.

A group of 49 construction companies formed themselves into the Nepal Biogas Promotion Group (NBPG) in 1995 (Karki *et al.*, 2007). NBPG has worked very closely with BSP to advertize biogas technology and encourage its use across the country. BSP and NBPG emphasized the value of biogas slurry as a fertilizer and liaised with the Nepal government organisation set up to promote renewable energies; the Alternative Energy Promotion Centre (AEPC), which had been set up by the Ministry of Science and Technology in 1996. By 2001, the BSP had installed 80,000 biogas plants (2,200 a year under BSP, compared with 750 a year previously).

In 2006, BSP was reorganized and became Biogas Sector Partnership (Bajgain, 2006). By the end of 2009, 205,820 biogas plants had been built under the scheme and there were 72 private biogas installation companies (Karki *et al.*, 2007). The number of biogas plants per head of population in Nepal was higher than anywhere else in the world. BSP had managed to register its plans for the period 2004 onwards with CDM (Clean Development Mechanism), set up by the Kyoto Protocol under UNFCCC (United Nations Framework Convention on Climate Change). This allowed BSP to continue with its subsidies without support from SNV or other aid organisations. The target under the CDM project was 135,000 plants between 2004 and 2010 (CDM, 2005) and almost 95,000 had been completed by the end of 2009. A change of the rules by CDM suggested that a reliance on carbon offset money for subsidies needed further consideration by BSP and AEPC (van Nes, 2007a). The Project Design Documents were rewritten to meet the new requirements (CDM, 2009) and were approved by UNFCCC. The total number of plants built under BSP by the end of 2011 was 250,476, over 0.25 million (Energy4All, 2012). BSP now comes under AEPC (Alternative Energy Promotion Centre), a Nepal government programme.

Biogas programmes in the rest of Asia

While there has been interest in biogas technology in other countries in Asia, with steady interest from Bangladesh and Sri Lanka, most of the projects have been much smaller than those in China, India and Nepal.

Bangladesh

The Bangladesh government has had an interest in biogas technology since the early 1970s based on the KVIC and other projects in India, but the interest has been primarily in academic institutions (Chaudhury, 2000). One project launched by the Environment Pollution Control Department (EPCD) in the early 1980s built 150 floating drum and 110 fixed-dome plants, but low-quality work and lack of follow-up caused most of them to fail. Other projects followed a similar pattern. The Bangladesh Council of Scientific and Industrial Research (BCSIR) sent staff for training in China in 1989 and started building plants to the Chinese design. Up to the end of 2007, 21,800 plants had been built with the help of subsidies from the government. In 1994, the government created the Infrastructure Development Company Ltd (IDCOL) with funding from the World Bank; the company had built 4,500 biogas plants by the end of 2005. A development bank, Grameen Shakti, which has been very successful with solar technology, started a building programme for biogas plants in 2004. SNV became involved with the IDCOL project in Bangladesh at the end of 2005 and aimed to repeat the success of BSP in Nepal (Bajgain, 2006). Over 10,000 plants had been built under this project up to the end of 2009 (IDCOL, 2009) and the number had increased to 20,756 by the end of 2011 (Energy4All, 2012).

Sri Lanka

Sri Lanka has had an ongoing biogas programme since the 1970s, but it has never become a mainstream technology in the way that it is seen in India. Many Sri Lankans are vegetarians, so the number of cattle and pigs is much lower than in India or China. This means that the biogas plants from India, based on animal dung, are less appropriate. The Energy Forum of Sri Lanka (Abeygunawardana *et al.*, 2003) and Practical Action (ITDG) tried to resurrect a biogas programme in 2000 (Munasinghe, 2000). In an initial survey they reported that only 33 per cent of the 5,000 plants that had been built were still functioning, and only 28.5 per cent were still being used. In 2003, a study of 350 plants built under previous programmes run by various groups showed that 76 per cent were still working (UNAPCAEM, 2002). The Sri Lankan programme is of interest as it developed a dry digestion system that used straw as the main feedstock in a batch digester (Bandara and Weerasinghe, 2000). SNV has done a feasibility study on starting a biogas extension programme in Sri Lanka (Ghimire, 2011) based on their Asia Biogas programme approach. Various other groups are also encouraging the use of biogas technology (HELP-O, 2013). However, groups need to adopt the right approach to enable biogas to find its place in Sri Lanka (De Alwis, 2000).

Vietnam

The Vietnamese programme has also been interesting in that it has used different technologies. When the country was split, North Vietnam was influenced by China, so people tried various early Chinese designs between 1964 and 1975, which were not successful. After reunification, interest was revived in biogas technology in 1981 and various groups became involved and built about 1,300 plants of both floating drum and fixed-dome designs (An, 2002). In 1990 various groups, such as the National Institute of Animal Husbandry (NIAH) of Vietnam and university research groups, worked together to develop the plastic-bag digester. This was seen as very much cheaper than the masonry design ($60 in 2000) and 20,000 units were installed between 1995 and 2005 (An, 2002). However, plastic-bag digesters were easily damaged and, even if they were well protected, had a life of two years or less. In 2000, an NGO, the Vietnam Gardening Association (VACVINA), which had been involved in installing plastic-bag digesters, developed a low-cost concrete model (Preston, 2002). In 2003 SNV set up a project in Vietnam, based on their work with BSP in Nepal, in which more reliable fixed-dome plants were installed with subsidies from aid organisations. Between 2003 and 2005, 12,000 plants were installed (van Nes, 2006); the project moved on to further stages, with a total of 123,714 plants installed by the end of 2011 (Energy4All, 2012). Cambodia has followed a similar path to Vietnam, only with a lag of three years. Only 400 plastic-bag digesters were installed. SNV expanded its work from Vietnam to Cambodia in 2005 (van Nes, 2006) and installed 14,972 plants by the end of 2011 (Energy4All, 2012).

The Philippines

In the Philippines, the emphasis has been more on industrial scale biogas plants. In 1972, a piggery operation with 10,000 pigs called Maya Farms was set up by Liberty Flour Mills, but it was quickly realized that there was a problem with the disposal of the slurry (Maramba, 1978). A biogas system was set up, which consisted of a series of batch biogas plants made of brick-lined pits in the ground. Each day one of the plants is emptied of exhausted slurry and refilled with fresh pig slurry. To process the slurry, 31 plants were built, one for each day that the slurry was retained. The gas from the plants was used to run a series of engines to generate electricity. Ten similar, but smaller, systems were set up in the Philippines by 1978. Small farm-scale plants were not seen as a priority in the Philippines (Barnett *et al.*, 1978), so the emphasis has been on agricultural and human waste disposal. More recently systems have been built that process waste from sugar-cane processing. Several systems that process dung from larger pig and cattle farms have been submitted to the CDM (van Nes, 2006). A feasibility study by SNV suggests that the market for small-scale domestic digesters is limited (Teune *et al.*, 2010) because there is a lack of animal dung and most people have a good supply of available firewood.

Thailand

The biogas programme in Thailand has followed a similar pattern to that in the Philippines. Small-scale biogas programmes have not developed, although some were started in 1975 (Barnett *et al.*, 1978). Fuel-wood has not been seen as a problem and fossil fuel has been easily available. Thailand has several large-scale biogas systems to process waste from large pig farms (Janssen, 2001) and from processing cassava (Carbon Bridge Pte Ltd, 2007). The interest in large-scale biogas systems seems to be growing fast, although government regulations need to be reconsidered to encourage energy from such renewable sources to be more effectively used (Prasertsan and Sajjakulnukit, 2006). An interesting dimension to the Thai programme is the supply of biogas to a village from a central plant based at a large pig farm (Limmeechokchai and Chawana, 2004). The work to encourage the building of small-scale biogas systems led to 2,300 plants being installed by the end of 2006 (Aggarangsi *et al.*, 2013). More recent interest has been in the construction of plants to process dung from herds of animals on much larger farms so that registration for CDM could be done more easily than for large numbers of small domestic biogas plants.

Indonesia

Indonesia has been involved in building biogas plants, with 6,000 plants installed by 2009 (van Nes *et al.*, 2009). Several universities were involved in research projects in the 1970s, which inspired the Ministry of Agriculture to build 200 pilot plants for farmers. The availability of firewood and subsidized kerosene meant there was a very low uptake for biogas technology until after the year 2000. Various groups, universities, NGOs and local and national government departments built biogas plants. Indonesia has several different designs in use: fixed-dome, floating drum (both steel and plastic) and plastic bag, and there is a lack of coordination between the various groups building the different designs. The Indonesian Biogas Programme (BIRU), supported by SNV and the Humanist Institute for Co-operation with Developing Countries (HIVOS), another Netherlands Development Agency, has had a programme to co-ordinate the work. In 2011 they claimed on its website (BIRU, 2011) that it was close to achieving its target of 10,000 plants by the end of 2013.

Korea and Taiwan

Korea was one of the countries in Asia that started an effective biogas programme, with 27,000 plants being built between 1969 and 1978 (Barnett *et al.*, 1978). However, the low winter temperatures that reduce gas production seem to have quenched people's enthusiasm for the technology, so the programme has not continued. In mid-2012, there was new interest in larger-scale plants that would generate biogas from food processing wastes, similar to those operating in the Philippines (Kim *et al.*, 2012).

Taiwan also launched an early biogas programme with 7,500 family sized plants built by 1975 (Barnett *et al.*, 1978). Taiwan developed the bag plant, using red mud plastic (a flexible type of PVC). As in Korea, enthusiasm for biogas technology faded, but there is a new interest in the technology (Tsai *et al.*, 2004).

Based on its initial success in Asia (van Nes, 2006), SNV is looking to set up new programmes in other countries. For example, in Myanmar, 1,200 plants had been built by 2012 (van Nes *et al.*, 2012). in Pakistan, they claimed to be on the way to building 14,000 plants by the end of 2013 (SEBCON, 2012).

Biogas programmes in Africa

The story of biogas in Africa has been much less positive. One of the early biogas pioneers was L. John Fry, who was a farmer in South Africa (Fry, 1975). However, other farmers in South Africa did not follow his example, and Fry relocated to USA.

Kenya and Tanzania

Tim Hutchinson (Ngigi *et al.*, 2007) built his first plant in Kenya in 1957. This was an above-ground plant based on a cylindrical steel tank. His company sold more than 160 biogas units between 1960 and 1986 in Kenya. The company is still building biogas plants, but at a much lower rate.

GTZ started a biogas programme in Kenya in the mid 1980s, building 250 biogas plants to the design made popular in India by KVIC, which use a floating steel-drum to store the gas (Kossmann *et al.*, 1997). Another 150 plants were built to the fixed-dome design once it had became more popular in India. The GTZ programme trained technicians and it was expected that they would be able to run small businesses installing biogas systems. About 2,000 plants were installed, but biogas technology gained a reputation for low reliability because there were many failures (Gitonga, 1997). There were several reasons for these failures, but the basic cause was a lack of quality control of the work of the technicians. There was little follow-up work, so customers did not know how to run and maintain their plants properly. People tried to build do-it-yourself plants rather than pay a trained technician to do it. They discovered that they lacked the important skills that the technicians had learned. People were also reluctant to handle the dung required to feed the plants, as it was considered a 'dirty' job (Gitonga, 1997; Ngigi *et al.*, 2007). Several of the technicians trained by GTZ have set up their own companies to build biogas plants, which are seen as good quality, but expensive. There is steady, if slow, progress with the technology in Kenya (Ashden, 2010).

The history of biogas in Tanzania followed a similar pattern to that in Kenya. The Small Industries Development Organisation built 120 floating drum plants between 1975 and 1984. The Arusha Appropriate Technology Project and the government-funded organization, Centre for Agricultural Mechanization and Rural Technology (CAMARTEC), built fixed-dome plants, as well as a locally developed design consisting of oil drums lashed together to make a floating gas holder (Kossmann *et al.*, 1997). Most of the work was in Arusha in the north of Tanzania. GTZ also became involved and offered training for the technicians involved in building plants. By 1992, there were 600 biogas plants in Tanzania, but the projects quickly lost momentum once the GTZ advisers had left.

Uganda

In Uganda, there has been an ongoing interest in biogas technology since 1970s, but very little has been achieved. SNV estimated that 500 to 600 plants were built since then (ter Heegde, 2009). There has been a lack of coordination between the various interested groups. In 2001, the Ministry of Energy and Mines Development wrote a plan to develop biomass energy, including biogas (MEMD, 2001). A local group, Integrated Rural Development Initiatives (IRDI), had been building biogas plants, but at a slow rate. The work is now being supported by aid organisations such as SNV and HIVOS under the Africa Biogas Partnership Programme and aims to make the work of biogas extension more effective.

Ethiopia

Another programme that was instigated by a national government was the Ethiopian Rural Energy Development and Promotion Centre (EREDPC), set up under the Ministry of Mines and Energy in 2002 (Boers, 2008). The first biogas plant was built in Ethiopia in 1979 and about 1,000 plants have been constructed since then, many by NGOs. A lack of coordination led to poor follow-up work, so many plants failed. SNV is working with local experts to develop a more effective programme (Boers, 2009).

Burundi and Rwanda

GTZ was involved in the Biogas Dissemination Programme in Burundi with the Ministry for Energy and Mining between 1984 and 1988 (Kossmann *et al.*, 1997). By 1992, 206 family-scale plants, and 84 institutional-sized plants had been constructed.

One of the engineers trained by GTZ in Arusha, Ainea Kimaro, went to work in Rwanda for the Centre for Innovations and Technology Transfer (CITT), part of Kigali Institute of Science and Technology (KIST). In 2001, he adapted the fixed-dome technology to install sewage systems for prisons in Rwanda. The country's troubled past had resulted in up to 10,000 men packed into prisons designed for a fraction of that number and the rather crude sewage systems had completely failed. KIST was able to build plants of up to 1,400 m^3 capacity that processed the sewage and generated enough gas to replace half the wood-fuel that was used to cook the food for the prisoners (Ashden, 2005). The Rwandan government funded the construction of these plants with help from the Red Cross. There were six such systems working or almost completed by 2005, with another three on order. The first one, built in Cyangugu in 2001, was working effectively and was being well maintained by the prisoners who had been trained in maintenance tasks under the supervision of the prison management. The sewage systems were underground and the prisoners had planted very attractive gardens in the soil that covered them.

Ghana

A project similar to that in Rwanda is being run by John Afari Idan in Ghana entitled Biogas Technology Africa Ltd (BTAL) (Idan, 2006). Dr Idan trained and gained experience as a sewage engineer in Holland and Germany. The design used by BTAL is very similar to that used by KIST, although the BTAL plants are usually smaller. These systems are built for institutions, such as hospitals, schools, universities, and hotels. The gas produced is used within the institution in their kitchens for cooking food. The first plant was built for a hospital in 1994 and BTAL completed 19 plants by mid 2008. Several more were being built, including one for the presidential palace.

SNV and HIVOS, with various other organizations, have set up Biogas for a Better Life (van Nes, 2007b), which seeks to develop biogas programmes across Africa (SNV, 2013), following the pattern established in Asia. Feasibility studies have been made in various countries in both East and West Africa, such as Ghana (Shrestha, 2008), Kenya (ETC Group, 2007) and several other countries.

Biogas programmes in Latin America

Various groups have been involved since the 1980s in setting up biogas projects in countries in Latin America and the Caribbean (Marchaim, 1992). For example, GTZ built several plants in Bolivia and Belize in the late 1980s (Kossmann *et al.*, 1997). However, few of these projects have been successful and none have expanded in the way that biogas projects have done in Asia.

There has been an interest in biogas technology in Brazil since the 1970s, but the programmes have not been very successful (Aparecida de Lima, 2006). In the 1980s, 200 biodigestors were installed in Paraíba State in the north east with the support of the Ministério das Minas eEnergia (MME). However, out of this total, only 4.6 per cent are still working and 96.9 per cent of the owners say that they have lost interest in the technology (Aparecida de Lima, 2006). The government has put pressure farmers with large-scale cattle and pig operations to process the animal's dung to reduce smell, pollution and methane emissions.

Since 2000, there has been an increasing interest in using biogas technology. The main emphasis has been on bag digesters: slurry lagoons covered by plastic sheets, with a working volume of 500 m^3 and above. There has been an ongoing interest in larger-scale plants for the processing of agro-wastes in the rest of Latin America, as well as in Brazil (Borzacconi *et al.*, 1995).

Smaller bag digesters, made of PVC, are being used in places such as Bolivia, Costa Rica, Ecuador, Mexico, Nicaragua, and Peru (Balasubramaniyam *et al.*, 2008; Garwood, 2010). Bag digesters are described in detail in Chapter 7 (Marti Herrero, 2008). These projects are fairly small; the large-scale programmes seen elsewhere have not been replicated in Latin America.

The way forward for biogas in the developing world

One feature of the successful biogas programmes has been the desire to share success more widely. The Chengdu Biogas Research Institute (Yao and Qian, 1992) has been running both national and international courses on biogas since it started. A total of 300 people from 70 different countries have attended the international training sessions (Liu and Zhao, 2008). GIZ has taken the fixed-dome idea from China and trained people in various countries in Africa to build it. SNV has set up projects in many countries, based on the success of BSP in Nepal. There is greater co-ordination of different groups, such as the Biogas for a Better Life Project. Such South to South technology transfer (with and without help from aid agencies) is expanding as communication between countries is becoming easier.

Biogas history in the developed world

The recent development of biogas technology in Europe and other developed countries has been well documented by groups such as Task 37 of IEA (International Energy Authority) (Wellinger and Murphy, 2013).

References

Abeygunawardana, A., Herath, W., Sokkanathan, K., Chandrasekera, B. and Bandara, S. (2003) *Energy and Poverty: Report of the Study on the Role of Energy in Poverty Alleviation, Colombo*, Sri Lanka: Centre for Poverty Analysis. <www.efsl.lk/reports/Report-FinaL CEPA STUDY.pdf> [accessed 17 July 2014].

Aggarangsi, P., Tippayawong, N., Moran, J.C. and Rerkkriangkrai, P. (2013) 'Overview of livestock biogas technology development and implementation in Thailand' *Energy for Sustainable Development* 17: 371–377 <http://dx.doi.org/10.1016/j.esd.2013.03.004> [accessed 17 July 2014].

de Alwis, A. (2000) 'Biogas - a review of Sri Lanka's performance with a renewable energy technology', *Energy for Sustainable Development* 6 (1) (March): 30-37 <http://dx.doi.org/10.1016/S0973-0826(08)60296-3> [accessed 17 July 2014].

An, B.X. (2002) *Biogas Technology in Developing Countries: Vietnam Case Study*, Ho Chi Minh City, Vietnam: MEKARN, SidaSAREC. <www.mekarn.org/procbiod/an.htm> [accessed 17 July 2014].

Aparecida de Lima, M. (2006) *Brazil Profile for Animal Waste Management*, Brazil: Brazilian Agricultural Research Corporation. <www.globalmethane. org/documents/ag_cap_brazil.pdf> [accessed 17 July 2014].

Ashden (2005) 'Kigali Institute of Science, Technology and Management (KIST), Rwanda', [website - case study] Ashden <www.ashden.org/winners/ kist05> [accessed 17 July 2014].

Ashden (2006a) 'Appropriate Rural Technology Institute: Compact digester for producing biogas from food waste', [website - case study] Ashden <www. ashden.org/winners/arti06> [accessed 17 July 2014].

Ashden (2006b) 'Shaanxi Mothers, China; Domestic biogas for cooking and lighting', [website - case study] Ashden <www.ashden.org/winners/shaanxi> [accessed 17 July 2014].

Ashden (2007a) 'Biotech: Management of domestic and municipal waste at source produces biogas for cooking and electricity generation', [website - case study] Ashden <www.ashden.org/winners/biotech> [accessed 17 July 2014].

Ashden (2007b) 'SKG Sangha, India; Biogas for cooking plus fertiliser from slurry', [website - case study] Ashden <www.ashden.org/winners/skgsangha> [accessed 17 July 2014].

Ashden (2010) 'Case study summary: Sky Link Innovators, Kenya', [website - case study] Ashden <www.ashden.org/winners/Skylink10> [accessed 17 July 2014].

Bajgain, S. (2006) *Implementation Plan: National Domestic Biogas and Manure Programme in Bangladesh*, SNV (Netherlands Development Organisation) <www.snvworld.org/download/publications/ndbmp_implementation_ plan_bangladesh_2006.pdf> [accessed 20 July 2014].

Balasubramaniyam, U., Zisengwe, L.S., Meriggi, N. and Buysman, E. (2008) *Biogas production in climates with long cold winters*, Wageningen, the Netherlands: Wageningen University <http://www.susana.org/docs_ccbk/ susana_download/2-855-new-study-prepared-for-wecfbiogasproduction-in-climates-with-long-cold-winters.pdf> [accessed 17 July 2014].

Bandara, H.M.C.K. and Weerasinghe, K.D.N. (2000), Faculty of Agriculture, University of Ruhuna, Sri Lanka <www.rshanthini.com/tmp/CP307/ ApplicationBiogasTechnology.pdf> [accessed 17 July 2014].

Barnett, A., Pyle, L. and Subramanian, S.K. (1978) *Biogas Technology in the Third World: A Multidisciplinary Review*, Ottawa: International Development Research Centre.

Bhat, P.R. and Chanakya, H.N. (2001) 'Biogas plant dissemination: success story of Sirsi, India', *Energy for Sustainable Development* 5, no <http://dx.doi. org/10.1016/S0973-0826(09)60019-3> [accessed 17 July 2014].

Biotech (2008) 'Biotech: Biogas plant consultancy, Kerala. Waste to energy projects.' [website] Biotech Ltd, Thiruvananthapuram, India <www.biotech-india.org/index.aspx> [accessed 17 July 2014].

BIRU (2011) 'Biogas Rumah: turn waste into benefits' [website] Indonesia Domestic Biogas Programme, Hivos (Humanist Institute for Co-operation with Developing Countries) <www.biru.or.id/en/> [accessed 17 July 2014].

Boers, W. (2008) *Biogas in Ethiopia: from Scepticism to Enthusiasm*, Ethiopia: SNV <www.snvworld.org/sites/www.snvworld.org/files/publications/biogas_in_ ethiopia_from_sceptic ism_to_enthusiasm_2008.pdf> [accessed 17 July 2014].

Boers, W. (2009) Progress Report National Biogas Programme Ethiopia, SNV, Ethiopia <www.snvworld.org/sites/www.snvworld.org/files/publications/progress_report_national_biogas_programme_2009_ethiopia.pdf> [accessed 17 July 2014].

Borzacconi, L. López, I. and Vinas, M. (1995) 'Application of anaerobic digestion to the treatment of agroindustrial effluents in Latin America', *Water Science and Technology* 32: 105–111 <http://dx.doi.org/10.1016/0273-1223(96)00144-8> [accessed 17 July 2014].

Breslin, W.R., Saubolle, B.R., Warpeha, P. and Leach, P. (1980) *3-Cubic Meter Biogas Plant*, VITA, Virginia, USA <www.builditsolar.com/Projects/BioFuel/VITABIOGAS3M.HTM> [accessed 17 July 2014].

van Buren, A., (1979) *A Chinese Biogas Manual*, London: Practical Action Publications (Intermediate Technology Publications), <http://developmentbookshop.com/chinesebiogas-manual-pb> [accessed 22 July 2014].

Carbon Bridge Pte Ltd (2007) *Univanich Lamthap Pome Biogas Project Thailand, Board*, CDM - Executive: UNFCCC Bonn, Germany <www.netinform.net/KE/files/pdf/Univanich Lamthap PDD AM22 May07v3.pdf> [accessed 17 July 2014].

CAAS (2013) 'Biogas Institute of the Ministry of Agriculture' (BIOMA, CAAS) [website - information] <www.caas.cn/en/administration/research_institutes/research_institutes_out_beijing/sichuan_chengdu/77948.shtml> [accessed 17 July 2014].

CDM (2005) *Biogas Support Program - Nepal (BSP-Nepal) Activity 1*, Board, CDM – Executive: UNFCCC, Bonn, Germany <http://cdm.unfccc.int/UserManagement/FileStorage/A4NYD8EXQY928HD61LHWHEIM82M BIN> [accessed 17 July 2014].

CDM (2012) *Project Design Document Form (cdm-ssc-pdd) - version 03 Updated,* Biogas Support Program - Nepal (BSP-Nepal) Activity-1 Board, CDM – Executive: UNFCCC, Bonn, Germany <https://cdm.unfccc.int/UserManagement/FileStorage/LB8249E3X610NRSZ5HTFKWVGOIM7CA> [accessed 17 July 2014].

Chaudhury, A.H. (2000) *Diffusion of Biogas Technology: a Community Based Approach*, Khulna University Studies 2 (1) (June): 7 <www.bdresearch.org/home/attachments/article/555/ChowdhuryAHBiogas.PDF> [accessed 17 July 2014].

Chauhan, D.S. and Srivastava, S.K. (2012) *Non-conventional Energy Resources*, New Age International, New Delhi. <www.newagepublishers.com/servlet/nagetbiblio?bno=001142> [accessed 17 July 2014].

Chen, L., Zhao, L., Ren, C. and Wang, F. (2012) 'The progress and prospects of rural biogas production in China', *Renewable Energy in China* 51: 58–63 <http://dx.doi.org/10.1016/j.enpol.2012.05.052> [accessed 17 July 2014].

Chen, Y. Yang, G. Sweeney, S. and Feng, Y. (2010) 'Household biogas use in rural China: A study of opportunities and constraints', *Renewable and Sustainable Energy Reviews* 14: 545–549 <http://dx.doi.org/10.1016/j.rser.2009.07.019> [accessed 17 July 2014].

Cheng, S. Du, X. Xing, J. Lucas, M. Shih, J. and Huba, E.-M. (2011) *4-In-1 Biogas Systems: a Field Study on Sanitation Aspects & Acceptance Issues in Chaoyang and Shenyang Municipalities, Liaoning Province*, Beijing, China: Centre for Sustainable Environmental Sanitation University of Science and Technology Beijing <www.ecosanres.org/pdf_files/4-in-1_

Household_Biogas_Project_Evaluation-20110620.pdf> [accessed 17 July 2014].

Cheng, S. Li, Z. Mang, H.-P. and Huba, E.-M. (2013) 'A review of prefabricated biogas digesters in China', *Renewable and Sustainable Energy Reviews* 28: 738–748 <http://dx.doi.org/10.1016/j.rser.2013.08.030> [accessed 17 July 2014].

Devkota, G.P. (2001) Biogas *Technology in Nepal: A Sustainable Source of Energy for Rural People*, Kathmandu: Mrs. Bindu Devkota, Maipee.

Devkota, G.P. (2007) *Renewable Energy Technologies in Nepal: An Overview*, Kathmandu: Universal Consultancy Services P. Ltd.

Energy4All (2012) *Brief progress and planning report the Working Group on Domestic Biogas*, SNV The Hague, Netherlands <www.hedon.info/tiki-download_item_attachment.php?attId=443> [accessed 17 July 2014].

ETC Group (2007) *Feasibility study (Kenya)*, ETC Group for Shell Foundation <www.snvworld.org/en/download/publications/feasibility_study_of_promoting_biogas_kenya_2007.pdf> [accessed 17 July 2014].

Fry, J. (1975) *Methane Digesters for Fuel Gas and Fertilizers*, John Fry and Richard Merrill.Santa Barbara, California <http://large.stanford.edu/courses/2010/ph240/cook2/docs/methane_digesters.pdf> [accessed 17 July 2014].

Fulford, D. (1988) *Running a Biogas Programme: A Handbook*, London: Practical Action Publications (Intermediate Technology Publications). <http://developmentbookshop.com/runninga-biogas-programme-pb> [accessed 22 July 2014].

Fulford, D. Poudal, T.R. and Roque, J. (1991) *Evaluation of on-going project: Financing and Construction of Biogas Plants*, United Nations Capital Development Fund. New York.

Garwood, A. (2010) *Network for Biodigesters in Latin America and the Caribbean: Case Studies and Future Recommendations*, Inter-American Development Bank; Sustainable Energy & Climate Change Unit.Washington DC < http://idbdocs.iadb.org/wsdocs/getdocument.aspx?docnum=35522800> [accessed 17 July 2014].

Ghimire, P.C. (2011) *Feasibility Study of National Domestic Biogas Programme in Sri Lanka*, People in Need, SNV Sri Lanka <www.snvworld.org/sites/www.snvworld.org/files/publications/biogas_feasibility_study_sri_lanka_2011.pdf> [accessed 17 July 2014].

Gitonga, S. (1997) *Biogas Promotion in Kenya*, Practical Action (ITDG) Publishing. Rugby, UK.

Harter, J. (2010) *The Diffusion of Biogas Technology and its Impact on Sustainable Rural Development in China*, Wolfen, Germany: Universität Duisburg-Essen.

ter Heegde, F. (2009) *Institutional arrangements for the Uganda Domestic Biogas Programme*, Uganda:SNV.<www.snvworld.org/en/Documents/Institutional_arrangements_for_the_Uganda_Domestic_Biogas_Programme_2009.pdf> [accessed 17 July 2014].

HELP-O (2013) 'HELP-O Newsletter-October 2013', [website - newsletter] HELP-O <http://helpo-srilanka.org/wp-content/uploads/2013/07/March-News-Letter-2013.pdf> [accessed 17 July 2014].

Idan, J. (2006) *Presentation on Integrated Sewage and Solid Organic Waste-to-Energy Project* [Presentation] BTAL (Biogas |Technology Africa Ltd).

<www.biogasonline.com/downloads/integrated_waste_to_energy_project.pdf> [accessed 17 July 2014].

IDCOL (2009) IDCOL: *Renewable Energy Projects,* Bangladesh: IDCOL (Infrastructure Development Company Limited). <www.idcol.org/brochure/Renewable_Energy_Initiative_of_IDCOL.pdf> [accessed 17 July 2014].

Janssen, P. (2001) 'Biogas Stops Thais From Turning Up Their Noses at Pig Farms', [website -journal] Digital Journal - Technology <www.digitaljournal.com/article/33412 > [accessed:18/07/2014].

Jiang, X., Sommer, S.G. and Christensen, K.V. (2011) 'A review of the biogas industry in China', *Sustainability of Biofuels* 39: 6073–6081 <http://dx.doi.org/10.1016/j.enpol.2011.07.007> [accessed 17 July 2014].

Kangmin, L. (2006) *Biogas China,* London: Institute of Science in Society. <www.isis. org.uk/BiogasChina.php> [accessed 17 July 2014].

Karki, A.B. (1996) *A Training Manual for Extension,* Kathmandu: CMS for Food and Agriculture Organization (FAO). <http://www.fao.org/3/a-ae897e.pdf> [accessed 17 July 2014].

Karki, A.B., Shakya, I., Dawadi, K.D. and Sharma, I. (Eds) (2007) *Biogas Sector in Nepal; Highlighting Historical Heights and Present Status,* Nepal Biogas Promotion Group (NBPG), Kathmandu, Nepal

Kim, Y.-S. Yoon, Y.-M. Kim, C.-H. and Giersdorf, J. (2012) 'Status of biogas technologies and policies in South Korea', *Renewable and Sustainable Energy Reviews* 16: 3430–3438 <http://dx.doi.org/10.1016/j.rser.2012.02.075> [accessed 17 July 2014].

Kossmann, W. Pönitz, U. Habermehl, S., Hoerz, T., Krämer, P. and Klingler, B. (1997) *Biogas Digest Volume IV: Country Reports,* Germany: ISAT and GTZ. <www.snvworld.org/en/publications/biogas-digest-biogas-country-reports-volume-iv> [accessed 17 July 2014].

Kurian, P.K. (2004) *Socio-Economic and Environmental Impact of Biogas Programme,* Thiruvananthapuram: Kerala Research Programme on Local Level Development, Centre For Development Studies. <www.cds.ac.in/krpcds/report/pkkurian.pdf> [accessed 17 July 2014].

Lawbuary, J. (2000) Biogas Technology in India: More than Gandhi's Dream? [website - report] Ganesha, www.ganesha.co.uk/Articles/Biogas%20Technology%20in%20India.htm [accessed 17 July 2014].

Limmeechokchai, B. and Chawana, S. (2004) 'Implication of Biogas Potential in Thailand: The Case Study of Livestock Farm'. 9-022 (P) The Joint International Conference on Sustainable Energy and Environment (SEE) Hua Hin, Thailand. <www.thaiscience.info/Article for ThaiScience/Article/3/Ts-3 implication of biogas potential in thailand case study of livestock farm.pdf> [accessed 22 July 2014]

Maramba, F.D. (1978) *Biogas and Waste Recycling: The Philippine experience,* Metro Manilla, Philippines: Maya Farms Division, Metro Flour Mills.

Marchaim, U. (1992) *Biogas processes for sustainable development,* FAO M-09. <www.fao.org/docrep/t0541e/t0541e00.htm> [accessed 17 July 2014].

Marti Herrero, J. (2008) *Low Cost Biodigesters to Produce Biogas and Natural Fertilizer from Organic Waste,* Innovation for Development and South-South Cooperation (IDEASS) Geneva, Switzerland. <www.ideassonline.org/public/pdf/BrochureBiodigestersENG.pdf> [accessed 17 July 2014].

MEMD (2007) *The Renewable Energy Policy for Uganda*, Ministry of Energy and Mineral Development of The Government of the Republic of Uganda. Kampala <http://www.rea.or.ug/index.php/policies-and-legislation? download=42:the-renewable-energypolicy-for-uganda.> [accessed 18 July 2014].

Munasinghe, S. (2000) *Biogas Technology and Integrated Development: Experiences from Sri Lanka*, Practical Action, Rugby, UK (ITDG). <practicalaction.org/media/view/6511> [accessed 18 July 2014].

Musafer, N.M. (2005) 'Biogas technology utilization in Sri Lanka', The International Seminar on Biogas Technology for Poverty Reduction and Sustainable Development, China: ESCAP. <unapcaem.org/Activities Files/A01/Biogas technology utilization in Sri Lanka.pdf> [accessed 18 July 2014].

van Nes, W. (2007a) 'Biogas for a better life: An African initiative', *Renewable Energy World* July 1 2007 [on-line journal] <www.renewableenergyworld.com/rea/news/article/2007/07/biogas-for-a-better-life-an-africaninitiative-51480> [accessed 17 July 2014].

van Nes, W. (2007b) 'Commercialisation and business development in the framework of the Asia Biogas Programme', Seminar on Policy options for expansion of community-driven energy service provision, Beijing, China.

van Nes, W., Aung, H., Soe, O. and Wah, S.L. (2012) *Feasibility of a National Programme on Domestic Biogas in Myanmar*, SNV and FAO, The Hague, Netherlands. <http://www.snvworld.org/download/publications/biogas_feasibility_study_myanmar_2012.pdf> [accessed 18 July 2014].

van Nes, W., Tumiwa, F. and Setyadi, I. (2009) *Feasibility of a national programme on domestic biogas in Indonesia*, SNV. Jakarta, Indonesias <www.snvworld.org/en/Documents/Feasibility_study_Indonesia_2009.pdf> [accessed 17 July 2014].

van Nes, W.J. (2006) 'Asia hits the gas: Biogas from anaerobic digestion rolls out across Asia' [on-line journal] *Renewable Energy World* Jan/Feb 2006 p102-111 <www.unapcaem.org/activities%20files/a01/asiahitsthegas.pdf> [accessed 18 July 2014].

Ngigi, A., Okello, B., Adoyo, F., Vleuten, F. van der; Muchena, F., Wilson, L. and Magermans, R. (2007) *Promoting Biogas Systems In Kenya: A Feasibility Study*, ETC Group for Shell Foundation, Nairobi. <www.snvworld.org/download/publications/feasibility_study_of_promoting_biogas_kenya_2007.pdf> [accessed 17 July 2014].

NIIR, B. (2004) *Handbook on Biogas and its Applications*, Delhi: National Institute of Industrial Research. <www.niir.org/books/book_pdf/114/niir-handbook-on-bio-gas-its-applications.pdf> [accessed 17 July 2014].

NSPRC (1985) *The Collection of Designs for Household Hydraulic Biogas Digesters in Rural A reas, Standardization*, China, State Bureau of: National Standard of the People's Republic of China.

Pal, S.P.; Bhatia, V.K.; Pal, B. and Routray, D. (2002) *Evaluation Study on National Project on Biogas Development*, New Delhi: Programme Evaluation Organisation, Planning Commission of India. <http://planningcommission.nic.in/reports/peoreport/peoevalu/peo_npbd.pdf> [accessed 17 July 2014].

Prasertsan, S. and Sajjakulnukit, B. (2006) 'Biomass and biogas energy in Thailand: Potential, opportunity and barriers' *SOUTH/SOUTH* 31: 599–610 <http://dx.doi.org/10.1016/j.renene.2005.08.005> [accessed 17 July 2014].

Preston, T.R. and Rodríguez, L. (2002) 'Low-cost biodigesters as the epicenter of ecological farming systems', [website - report] Ho Chi Minh City, Vietnam: MEKARN, SidaSAREC. <www.mekarn.org/procbiod/prest.htm> [accessed 17 July 2014].

Saubolle, B.R. and Bachmann, A. (1983) *Fuel Gas from Cowdung Kathmandu*, Kathmandu, Nepal: Sahayogi Press

SEBCON (2012) *Biogas User's Survey 2011*, Pakistan Domestic Biogas Programme and SNV Islamabad <http://www.snvworld.org/download/publications/biogas_user_survey_pakistan_2011.pdf> [accessed: 18/07/2014].

Shah, O.P. (2006) *Sustainable Waste Processing in Mumbai: Using the Nisargruna Technology*, MSc Thesis. Borås, Sweden: Hogskolan i Borås. <http://www.kingdombio.co.uk/ShahThesis.pdf> [accessed 17 July 2014].

Shrestha, R. (2008) *Physical Feasibility of Domestic Biogas in the Upper East Region of Ghana*, Accra, Ghana: SNV. <www.snvworld.org/download/publications/ndbp_physical_feasibility_study_ghana_2008.pdf> [accessed 17 July 2014].

Singh, R.B. (1974) *Bio-gas Plant: Generating Methane from Organic Wastes*, Ajitmal, India, PRAD

SNV (2013) *SNV in Africa: People, Partnership, Progress*, SNV., The Hague, Netherlands<www.snvworld.org/sites/www.snvworld.org/files/publications/snv_people_partnership_progress_2013.pdf> [accessed 17 July 2014].

Teune, B., Orprecio, J., Dalusung, A. and Yeneza, G. (2010) *Feasibility Study of a National Biogas Programme on Domestic Biogas In The Philippines*, SNV and Winrock International.The Hague, Netherlands<www.snvworld.org/sites/www.snvworld.org/files/publications/feasibility_study_of_a_national_ domestic_ biogas_programme_the_philippines_2010.pdf> [accessed 17 July 2014].

Tsai, W.T., Chou, Y.H. and Chang, Y.M. (2004) 'Progress in energy utilization from agrowastes in Taiwan' *Renewable and Sustainable Energy Reviews* 8: 461–481 <http://dx.doi.org/10.1016/j.rser.2003.12.008> [accessed: 18/07/2014].

Tu, Q. (2010) *Biogas Use in China and its Sustainability* [Presentation], Chinese Academy of Social Sciences, Wageningen University, Netherlands <www.wageningenur.nl/upload_mm/9/5/b/3bed97aaff6c-4e23-ae31-759bf2fd485a_Qin_Tu_BiogasSustainabilityinChina.pdf> [accessed 18 July 2014].

Vasudeo, G. (1993) *Biogas: a Manual on Repair and Maintenance*, Kanyakumari, India: Vivekananda KendraKendra (VK NARDEP) <http://www.vknardep.org/publications/english-books/197-biogas--a-manual-on-repair-a-maintenance.pdf> [accessed 29 August 2014].

Wellinger, A. and Murphy, J. (2013) *Biogas Handbook: Science, Production and Application*, Cambridge: D Woodhead Publishing Series in Energy. <www.iea-biogas.net/biogashandbook.html> [accessed 17 July 2014].

Wilson, P. (2013) *The Generation of biogas On-farm Using Animal and Dairy Waste*, Melbourne: International Specialised Skills Institute. <www.agrifoodskills.net.au/resource/resmgr/fellowship_reports/wilson_report_(lowres).pdf> [accessed 18 July 2014].

Wu, L. (2003) *Up-to-date Status of Biogas Industry Development in China*, Chengdu: China Biogas Society. <www.worldenergy.org/documents/congresspapers/libinw0904.pdf.> [accessed 17 July 2014].

Yao, Y. and Qian, Y. (1992) *The Biogas Technology in China: Chengdu Biogas Research Institute of the Ministry of Agriculture*, Beijing P.R.C.: Agricultural Pub. House. <www.worldcat.org/title/biogas-technology-in-china/oclc/826826537> [accessed 17 July 2014].

Zuzhang, X. (2013) *Domestic Biogas in a Changing China: Can Biogas Still Meet the Energy Needs of China's Rural Households?*, London: International Institute for Environment and Development. <http://pubs.iied.org/16553IIED.html> [accessed 18 July 2014].

CHAPTER 3

Aspects of a biogas programme

Abstract

The process of making biogas technology available widely in a country involves a large number of different factors. The provision of domestic biogas units to very large numbers of people, such as in China, India, Nepal, and elsewhere, requires careful planning and there are many challenges that face people who work on such a project. There are many benefits of biogas technology, such as replacing firewood or LPG. Biogas plants can be made to a range of different scales, from small backyard systems to large industrial systems. Particular challenges are faced in economic, organizational, and research and development areas.

Keywords: biogas, anaerobic digestion, aspects

Challenges of a biogas programme

The process of making biogas from biomass materials is microbiological. The biochemistry of the process has been studied, but it is very complex and the results of these studies offer few insights into ways in which the efficiency and effectiveness of the process can be improved. The technical aspects of biogas technology involve making a cheap and reliable container in which the mixture of food wastes and bacteria can be held in the form of slurry with a means of collecting the gas produced. Equipment to use the gas for the required purposes also needs to be available (Fulford, 1988). This may be in the form of cooking stoves, if the biogas is to be used for domestic purposes, or suitable engines and generators, if it is to be used to generate electricity.

The appropriate feedstocks need to be available in the correct quantities. The early biogas programmes relied primarily on animal dung as the main feed material, but there is a growing interest in the use of food and other agricultural residues. The plants built by the early programmes tended to be too large for the supply of feedstock and people had too high an expectation of the amount of gas that could be produced (Basudev, 2008).

The economics of biogas is critical; people will only pay for a solution if they feel they are saving or earning enough to justify the cost. Biogas does carry a 'green incentive'; people will pay extra if they can feel good about saving the environment, but only up to a point. The most effective programmes have been ones with subsidy, with the justification for the subsidy being that the country as a whole is paying to reduce the damage done to the environment by an overuse of fuel-wood. More recently, the issue of climate change means

http://dx.doi.org/10.3362/9781780448497.003

that people are willing to provide finance for the replacement of fossil fuels by an effective source of renewable energy (carbon offset finance).

The funding for the programme is important, especially for loans to allow people to buy the plants. As with most renewable energy technologies, there is a high front-end cost that leads to a much lower running cost. This means that people need to take out a loan, which can be repaid from the savings and earnings that arise from using the technology. In both Nepal and India, the involvement of local rural banks and micro-finance institutions (MFIs) was essential for the growth of biogas projects.

There are social implications to the technology. In India and Nepal, biogas is often called *'gobar gas'* as the word *'gobar'* is the word for cow dung. In Hindu culture, a cow is a holy animal, so its dung is a very acceptable source of cooking fuel. Pig dung and human faeces are much less acceptable (the biogas produced is almost identical), although people are overcoming their reluctance to link a latrine to their biogas plant. In China, there has been a long tradition of composting residues, so biogas plants were seen as an extension of that tradition with the added benefit of producing cooking gas. The emphasis in the early Chinese programme was more on the smell-free compost produced than on the gas. These social factors go a long way to explain the successes of the programmes in these countries.

Political dimensions can help or confuse a programme. The Indian programme was directed centrally, by the Khadi and Villages Industries Commission (KVIC) and then by the Ministry of Non-conventional Energy Sources (MNES). Civil servants set targets that often distorted the operation of the programme in the rural areas. In China, the biogas programme was accelerated too quickly by Chairman Mao and his Great Leap Forward. Many plants built under the early programme were of poor quality and failed after a short time.

Publicity can be used to encourage a biogas programme. People are much more likely to accept a new idea if they have heard about it from the media. All of the effective programmes have worked with journalists and science writers to make sure that articles appeared in newspapers, on the radio, and on TV that informed people of biogas technology and its environmental benefits. The best publicity for such a programme is word-of-mouth; once one person in an area has a plant, many of their neighbours see the benefits and want one as well.

When a government department, an aid agency, or a commercial organization plans to start a biogas programme, it needs to take all these aspects into consideration. This means that a biogas team should include people with expertise in a range of fields (van Nes *et al.*, 2009). The type of organization that leads the extension programme does not seem to affect its effectiveness: aid organizations, government agencies, NGOs, and commercial companies have all successfully run biogas programmes. However, there is a need for different organizations to work closely together for a programme to work well: the installation group needs strong and close support from aid groups, national and local governments, banks and micro-finance groups, academic and other research institutions, and the local media to ensure the programme runs well.

Benefits of biogas technology

The reason that rural, domestic biogas programmes have grown so fast in many countries is that biogas offers a wide range of recognisable benefits to the users, the local environment, and the wider environment. The economic value of these benefits has been difficult to assess and many economic analyses have led to inconclusive results. When rural, domestic biogas users become experienced in using their plants, they do discover the real value of these benefits.

The immediate benefit from using a biogas plant is a steady supply of clean cooking fuel (see front cover), which replaces other fuels such as firewood, kerosene, or LPG. Although many governments do subsidise kerosene and LPG, the economic benefits of biogas can be seen immediately. Both kerosene and LPG are very difficult to obtain in more remote rural areas and there is a cost for the transport of these fuels, as well as for their purchase.

If biogas replaces firewood, the cost benefits are less obvious, but there are many other improvements to people's daily lives. Wood smoke causes irritation to the nose, throat, lungs, and eyes and the use of biogas removes this immediately. In many places, agricultural residue and dried dung is used alongside wood as domestic fuel. These fuels produce even greater amounts of smoke than wood and if they are burnt, the potential value of their use as compost to put on crops is lost.

Biogas is instantly available in the morning and heat can be easily controlled by turning a valve. The children of a biogas plant owner can be given a hot breakfast before they go to school. The outsides of cooking pots are very clean, without soot. Women's clothes and skin are also much cleaner. The removal of smoke has very positive health benefits, particularly for women and children (AQP, 2004; Smith, 1993).

The incidence of burns to children can be very high in areas where people use open fires for cooking because children can easily fall into a fire. Open fires can also spread to the rest of the house if the user is careless. In some areas, house fires are very common. A biogas stove is much less dangerous because it is contained. Although fuelling a biogas plant requires the collection of animal dung, firewood collection takes much longer. Carrying loads of firewood is hard work and can strain people's muscles. Dung for a biogas plant needs mixing with water. Wood requires cutting and storing in a dry place. User surveys have shown that people using biogas save over three hours a day, which can be used for productive activities (Ghimire *et al.*, 1996).

User surveys in Indo–China suggest that the main benefit of a biogas plant is the reduction of smell from the use of dung from pigsties close to the house (Dung *et al.*, 2009). People can allow piglets to grow larger before they are sold, giving a higher price. Without a biogas plant, larger pigs produce more dung and therefore more smell. With a biogas plant, the smell is removed from the larger amounts of dung and more biogas is produced.

Biogas can also be used for lights to replace kerosene, although the efficiency of biogas for lighting is not more than for kerosene. Kerosene smells and can spill and catch fire if lamps are knocked over. Gaslights are fixed, so the danger of fire is reduced. User surveys show that a biogas plant can save a family about 32 litres of kerosene a year, if biogas replaces kerosene for lighting (Bajgain and Shakya, 2005).

Many biogas plants have latrines attached, so their installation improves sanitation (Figure 3.1). Pathogens and parasites are not completely eliminated in a biogas plant, but they are severely reduced. A representative from World Wildlife Fund (WWF) claimed that they had found that a latrine could reduce the number of people bitten by snakes. Traditionally women hide behind bushes at dawn and dusk for toiletry purposes and are at risk of being bitten by the snakes that spend the night there.

The effluent slurry from a biogas plant has an improved fertilizer value over the use of raw dung. The fertilizer value can be increased by further processing, such as by mixing it with dry biomass material and composting it. Composting also further reduces pathogens present in the slurry.

Average values of the benefits were estimated for an average family (of six people) and are presented in the following table (Bajgain and Shakya, 2005).

There are benefits of biogas for the local environment, as the main use of biogas technology is the replacement of wood for cooking fuel, so less firewood needs to be collected. Discussions with Santosh Mani Nepal, the Director of WWF in Nepal (Fulford *et al*, 2012), suggested that a family with a biogas plant in Nepal saves about 2,000 kg of wood a year. He had calculated that the installation of 1,000 biogas plants in areas around wildlife parks in

Figure 3.1 Woman by latrine attached to biogas plant

Table 3.1 Estimated benefits from the installation of a domestic rural biogas plant

Benefit	Values for an average biogas plant
Reduction of workload (especially women)	900 hours per year (2.5 hours per day)
Saving of firewood	1,800 kg per year
Saving of agricultural waste	600 kg per year
Saving of dried dung	250 kg per year
Saving of kerosene	45 litres per year
Reduction of CO_2 emission	4.5 ton per year
Improvement of health	No indoor smoke pollution, improved sanitation
Increase of agricultural production	Increase (up to 40%) in yields

Source: (Bajgain and Shakya, 2005)
Note: the values were collected from a range of sources and average values taken

the Terai (plains) area of Nepal could save 33.8 ha of forest from clear felling. Tree cover is essential for the survival of many animal species, such as tigers (Roberts, 2011), and reduces the degradation of land. Deforestation is seen as a major cause of landslides in Nepal, because tree roots stabilize the ground on steep hillsides. Trees also allow rain to penetrate into the soil along their branches, trunk and roots, maintaining the watersheds that store water and supply springs.

The use of a local supply of fertilizer from biogas slurry reduces the need for artificial fertilizers, which can pollute the environment. If bioslurry from a biogas plant is carefully composted with dry biomass, its quality can be further improved. Good-quality compost acts to enhance the soil because it helps the soil to retain nitrogen and other plant nutrients in the top layer. The compost releases plant nutrients slowly, so the plants can use them. Inorganic chemical fertilizers, such as urea, are water soluble and leach out of the soil more quickly. Artificial fertilizers pollute watercourses and encourage the growth of algae and other weeds, which take oxygen from the water and cause fish and other water creatures to suffocate and die.

Another effect of deforestation is the removal of a carbon sink from the wider climate system. Trees absorb carbon dioxide from the atmosphere, so their removal reduces the rate at which carbon is sequestered. If they are used unsustainably as wood fuel and are not replanted, they also add carbon dioxide to the atmosphere.

The use of fossil fuels also adds carbon to the atmosphere. When fossil hydrocarbons are extracted, converted into fuels and burnt, the carbon that was locked underground is converted into extra carbon dioxide in the atmosphere. These two processes are recognized as contributing to climate change, because the balance of heat transfer from the Sun to the Earth is affected by the amount of carbon dioxide and other trace gases in the atmosphere.

The use of biogas plants reduces the rate of deforestation and also replaces the use of kerosene for lighting, therefore it has been recognized that a biogas extension project does reduce the amount of carbon dioxide released into the atmosphere (Devkota, 2003). A methodology has been defined by CDM (Clean Development Mechanism) to assess how much fossil carbon is being offset by constructing biogas plants to replace wood fuel and kerosene (CDM, 2012). Further refinement of the CDM methodologies have been able to include the use of the fertilizer from composted biogas slurry (ter Heegde, 2005). The production of inorganic fertilizers is very energy intensive, so its replacement reduces the use of fossil carbon.

Scale of biogas systems

The domestic plants that have been built in huge numbers, especially in China, India, and Nepal, are typically small enough to fit in a small garden or farmyard. The gas is primarily used to supply cooking fuel, although many people like to use surplus gas for running gaslights. The working volume of a domestic plant is typically 4 m^3, although most programmes supply a range of sizes from 1 m^3 to 20 m^3. The rate of gas production is affected by temperature, so in cooler climates plants are made with larger volumes to compensate for lower gas generation rates. Some programmes quote the plant size as the amount of biogas that can be generated in one day, but because this is affected by temperature, it is not a reliable way to define the size of a plant. The working volume is a measure of the space inside a biogas plant that contains the slurry that generates biogas.

Biogas plants that utilize animal dung tend to be larger, because the available material for digestion in the dung is less, since the animal has already used the most easily digestible food matter. The dung from four cattle (40 kg of wet dung) or eight pigs is usually quoted as sufficient to provide a small family with all its cooking fuel. However, there are examples of people using the dung from one cow or two or three pigs to provide sufficient gas to cook for a small family, with very careful management.

The plants that have been built in the largest numbers are mainly underground, which is an advantage where space is in short supply. In China, the biogas plant is often built under the pig shed to save space. Since the effluent is a good fertilizer, people often grow flower or vegetable gardens around and over the top of plants. The concrete dome that forms the roof of an underground plant needs to be strong enough to prevent cracking, so it can easily take the weight of people or animals walking over the back-fill earth. It would be unwise to place a biogas plant under a roadway over which vehicles will pass.

Biogas plants that utilize food wastes can be smaller. A domestic food-waste plant, as made in India, has a working volume of 1 m^3. It is supplied with about 1 kg of dry matter each day, although this is usually diluted with several litres of water to form slurry that can flow down the inlet pipe. If a family uses only its own food wastes, they can save between 25 and 50 per cent of the cooking fuel (usually LPG or kerosene) that they previously used

(Ashden, 2007a). Many people with such plants obtain extra food wastes from restaurants or other commercial food processors, so they can replace all of the fossil fuel they use for cooking (Ashden, 2006). The attraction of food-waste digesters is that they can be used in urban and suburban areas.

If biogas is to be used to supply an engine with fuel, the plant size needs to be larger. A typical dual-fuel generator (using 80 per cent biogas and 20 per cent diesel) of 5 kW electrical output needs about 3 m^3 per hour of biogas (NIIR, 2004), the same amount of gas that a typical domestic plant produces in a day. To make generating electricity worthwhile, a biogas plant at least five times bigger than a typical domestic plant is required. Such a plant would require the dung from about 30 cattle (300 kg of wet dung) or 60 pigs, or at least 5 kg of dry foodwaste a day to run.

In India and Nepal, these larger plants have been used to provide irrigation pumping or food milling. A commercial dairy or animal-fattening unit, with sufficient dung, can use such a system to supply an electrical generator with fuel to deliver its energy needs. Larger systems have also been run as cooperatives, with several families working together to supply the dung. These schemes have proved technically successful, although social factors have often caused problems. Typically larger systems use a floating drum plant with a 30 m^3 or 40 m^3 working volume.

Similar larger-sized plants have been used to process food and sewage wastes in Kerala (Ashden, 2007b). Institutions, such as schools and hostels, can replace 50 per cent of their LPG or kerosene use for cooking from the waste products of the people who attend the institution. Larger fixed-dome underground plants (working volumes from 50 m^3 to 1,400 m^3) have been used to process the sewage from institutions in Nepal (Lohri *et al.*, 2010), Rwanda (Ashden, 2005a), and Ghana (Idan, 2006). These include prisons, hospitals, schools, universities, and hotels. Again, the main use of the gas is for cooking and half of the previously used cooking fuel (wood, LPG or kerosene) has been saved.

In the developed world, biogas has been primarily seen as a large-scale process. Large central plants are used, to which dung, food wastes, or agricultural crops grown to produce energy are transported (Wellinger and Murphy, 2013). For example, in the German programme, the electricity generated from biogas plants ranges from less than 150 kW up to 20 MW per unit. A 500 kW plant uses a digester tank of 15,000 m^3 working volume and uses about 1,000 tonnes of biomass material a day. Most systems on mainland Europe use several tanks of this size, or sometimes smaller. There is usually an element of redundancy, so one of the tanks can be emptied and cleaned, while the rest continue to generate biogas. The tendency in Europe was to go for larger and larger systems for economies of scale, such as the 20 MW unit that was constructed in Northern Germany in 2008 (EnviTec, 2008). More recently, sizes have been smaller, as less material has to be transported to sites (EBA, 2013).

This large-scale approach has been exported to India and China (Kotrba, 2007), where it is being used with food-processing wastes such as brewery effluents and sugar residues. German industry has been very effective in exporting this technology to many other parts of the world, including the Philippines, Thailand, Taiwan, and Brazil.

There is an apparent gap in the technologies available for biogas between the small-scale domestic systems used in huge numbers in Asia and the large-scale systems that the German industry is enthusiastically exporting. The largest domestic system is considered to be 20 m^3, while the smallest economically viable large-scale unit is seen as 10,000 m^3.

The floating drum systems used in India for processing food wastes from municipal sources can be as large as 70 m^3 (Devkota, 2001), but the size is limited by the difficulty of transporting large metal drums. The answer in China was the Puxin digester (Puxin, 2013), which was made as a group of 10 m^3 units all linked together to make 100 m^3 or 200 m^3 units. However, they have also developed a unit of 100 m^3 that is cast from concrete using moulds. The most effective larger sized low-cost units are made in Africa (Rwanda, Ghana and Tanzania), where single fixed-dome plants have been made as large as 100 m^3 (Ashden, 2005b). These 100 m^3-sized units have been linked together to form digesters with a total volume as large as 1,400 m^3.

Economic challenges

Many economic analyses have been made of biogas technology over the years (Barnett *et al.*, 1978; Kossmann *et al.*, 1999) but the results were usually inconclusive. However, more recent analyses look more positive (Dhakal, 2008; Karki *et al.*, 2005). The running costs of a biogas plant are low, especially if the feed material has low or negative costs (i.e. payment has to be made to remove the waste material), but the capital cost per unit of energy generated is high. This is common to most renewable energy technologies, mainly because the energy density in biomass, wind or sunlight is much lower than that in coal, oil, gas or even hydropower. The physical dimensions of a plant to extract that energy must be larger per unit of energy produced than for a conventional energy plant.

The need to spread the capital cost of the plant over its lifetime using time depreciated cost–benefit analysis depends on various economic assumptions that favour conventional systems. Conventional plants rely on economies of scale to extract large amounts of energy in centralized units. Renewable energy technologies tend to be diffuse; plants are small and scattered, which increases the capital cost per unit of power generated. However, the economics of centralized power systems often do not take the cost of power distribution into account, because this cost is often funded by a different route. Such economic analyses also ignore the cost of the disposal of wastes from the operation, which can be very significant especially for nuclear power.

Most biogas systems in the world are used to supply domestic fuel, to replace firewood, kerosene or LPG for cooking, and sometimes for lighting. The cash value of firewood in the market is very low, because poorer people collect it and they have little ability to influence the price they receive. The actual cash value of the wood is often zero, because farmers will collect it themselves from the forest, even though this is illegal in many places. The analyses usually show that the cost–benefit results are very sensitive to the price of the fuel the biogas replaces. Even where biogas replaces kerosene or LPG, which have a market price, these fuels are often subsidised by governments, so the price paid by the user is often much less than the real market price.

The use of fuel-wood can have high environmental costs if wood is removed unsustainably from steep hillsides and from watershed areas. Deforestation can result in land erosion, landslides and floods, giving rise to irreversible changes to a local landscape. These environmental costs are not included in the price of the fuel in the market. The use of fossil fuels is seen to have an environmental impact that is even wider because they release carbon into the atmosphere adding to the greenhouse effect provided by carbon dioxide and other gases. If animal dung and food wastes rot down naturally in anaerobic conditions they can produce methane. Methane is seen as being over 20 times more potent as a greenhouse gas than carbon dioxide (estimates range from 15 to 33 more potent, depending on the assumptions used).

Biogas is seen as a carbon neutral technology because the methane produced is derived from carbon that has been removed from the atmosphere by growing plants. If this methane is burned in a cook stove or an engine, the carbon dioxide it produces only replaces that taken in by the plants when they grew. As long as the plant material is harvested sustainably, i.e. the grass eaten by the animals continues to grow, or the food planting/harvesting cycle continues, then this argument is valid. People have argued that, since the compost generated as a by-product of a biogas system enables plants to grow more vigorously, a biogas plant can actually remove carbon from the atmosphere. Also some of the carbon left in the compost will remain in the soil, acting as a soil conditioner.

These arguments have encouraged governments in China, India and Nepal to provide subsidises for biogas technology. Typical subsidies have been 25 per cent to 50 per cent of the capital cost, although up to 75 per cent has been provided to poor villagers in very environmentally sensitive areas. Even with these subsidies, it was estimated that only 10 per cent of the population in India could afford domestic biogas units (Tam, 1982). The provision of loans by Grameen (rural) banks and micro-finance institutions has enabled the steady growth of the programmes in India and Nepal.

More recently, the setting up the United Nations Framework Convention on Climate Change (UNFCCC) and the adoption of the Kyoto protocol in 1997 has enabled some biogas projects to claim carbon offset finance. Obtaining CDM finance has proved to be a complex process (ter Heegde, 2008). BSP Nepal (BSP) in Nepal, with the support of the Netherlands Development Organisation (SNV) was able to get the first CDM project document approved in 2004. A change in the application process by UNFCCC meant that BSP had to write a new set of documents in 2010 (CDM, 2012), but this has been a useful source of extra finance for the programme (Stone, 2012). Biogas programmes in other countries, such as Vietnam, have also benefited from carbon-offset finance (ter Heegde, 2005).

Although an agreement to continue the Kyoto Protocol received commitment from fewer countries, there is a continuing interest in carbon-offset finance. The CDM mechanism is being continued (CDM, 2013) despite its complexity (Zurbrügg and Volkart, 2012) and there is also a growing market in voluntary offsets. A range of organizations accept money from companies who want to demonstrate their green credentials and use it to offer subsidies for projects such as those that install biogas (the voluntary carbon market). Various projects in India have signed such agreements (MyClimate, 2012). From the mid 2000s, such agreements started replacing government subsidies as the main way to reduce the installation cost for biogas technology for poorer customers.

The economics of biogas systems used to run engines to generate electricity has also been analysed. Engines require larger biogas plants and adequate quantities of feed material. Equipment is usually provided to perform tasks such as handling the feed material, stirring the digester, and controlling the engine. This means that the capital cost of such a system per unit of energy generated is very much higher than for a domestic system, but the return on the investment is also much higher because the energy generated can be sold, usually as electricity, into the local grid. However, similar economic analyses to those on domestic biogas give similar results; biogas offers marginal economic returns on investment. An analysis done in the USA (Ross and Drake, 1996) in 1996 shows an IRR (internal rate of return) of 6.3 per cent, which is fairly typical.

This argument encouraged the German government to offer subsidised feed-in tariffs for on-farm electricity generation to make biogas systems look more economical in 2001 (Kram, 2007). The scheme was revised in 2004, to provide higher tariffs for electricity generated from energy crops, as opposed to food wastes and agricultural residues. Similar arguments were used to justify these subsidies as were used to justify those in Nepal, India and China. European governments had strongly supported energy companies to build generating plants in the past, so new renewable energy technologies needed similar support. In reality, much of the conventional electricity generating structure had been built in Europe when electricity supply was under government owned corporations, before privatization.

Organizational challenges

The range of different aspects involved in running any programme related to renewable energy means getting people from different disciplines working together. The people who become first involved tend to be those who are inspired by the technology; renewable energy offers different types of technical challenges compared to those of conventional energy systems. In order to get a programme working effectively, many other types of expertise are also required, such as microfinance, agricultural extension, mass communication, sociology, and politics. People from these different disciplines need to be able to communicate with each other for such a programme to work (van Nes *et al.*, 2009).

Biogas covers a wide range of different technologies, ranging from microbiology to structural engineering, so even the technical staff need to learn to communicate across different subject barriers. Anaerobic digestion is used in different applications, such as sanitation, the clean up of commercial wastewaters and, more recently, the disposal of food wastes, so there is a further set of subject barriers across which to communicate.

The managers of a biogas programme therefore need to understand the wide range of aspects that such a programme covers and to communicate with the specialists in the different areas. The technicians who are involved in the start of such a programme must be ready to see these wider challenges and include them in their plans.

The need for well-trained staff is crucial for the success of a biogas programme. The technical staff who do the work of building the plants must maintain a high standard of quality in their construction work. The failure of one plant due to poor construction can quickly reduce the reputation of the whole programme. The most effective way to train such staff is on-the-job, although the best people to hire are those who have a good previous experience in building construction (Karki, 1996).

A closely related organizational requirement is for quality control of the plants that are built. The work of the technical staff needs to be regularly inspected by an outside inspector to ensure that their work is always up to standard. If outside agencies are involved, such as financial institutions offering loans and/or subsidies, there is a good incentive to ensure the plants for which finance is offered will continue to function effectively. Such groups can impose financial pressure if they do not release funds until the plants, or a reasonable sample, have been inspected for the quality of the workmanship (Fulford *et al.*, 2012). In order to check the quality of biogas plants and to ensure they continue to function, another aspect is that of follow-up. The internal supervisors need to revisit the plants they have built to check that the customers are using them correctly. Outside inspectors need to visit a sample of the plants that have been built to check that this is happening.

BSP was set up by SNV primarily to offer training and quality control for the Nepal biogas programme. The work of building biogas plants was offered to a number of small, local construction companies that were willing to have their staff trained. The programme also offered loans and subsides through their relationship with local banks and microfinance institutes, therefore BSP was able to place pressure on these companies to maintain high standards (de Castro *et al.*, 2008). The result has been the very high success rate of 97 per cent of plants still working five years after they were built (Ashden, 2005c).

As well as good quality-control a second concept is that of a guarantee to the customer. The Gobar Gas Company (GGC) in Nepal, which was the only organisation building biogas plants before BSP was set up, did guarantee that their plants would work for at least one year. If a plant did fail, due to poor construction, the company agreed to rebuild it. This approach ensured that company staff kept to high standards and formed the basis of the BSP quality-control system.

SNV has taken this model and used it in many other countries, such as Bangladesh and Vietnam. It has adapted the model according to local conditions and the organizations with which it is working (van Nes *et al.*, 2009).

SKG Sangha operate a similar quality control approach, although it is internal to the organization. Individual technicians are made responsible for their work and some of the payment is withheld for six months, so that the quality of the work can be checked. There is also a promise given to the customer that any defect in workmanship will be corrected. The promise is not called a 'guarantee', as other programmes in India, which claim to offer guarantees, have misused the term and do not keep their promises. In order for these checks to be made, it is important to keep a good record of each plant built, the customer for whom it is made and which technicians build each plant.

Research and development

The two main biogas projects in China and India both followed several decades of independent research and development (Chen *et al.*, 2010; Lawbuary, 2000). Although the project run by Development and Consulting Services (DCS) could learn from the work done in both India and China, a major part of the project funded by United States Agency for International Development (USAID) was research and development to find the most appropriate ways in which the technology could be used in Nepal. In the same way that the subject areas required for biogas extension involve a wide range of subject areas, the work of R&D also requires researchers from a range of disciplines. An effective extension programme requires the resources of people who have a good understanding of the many different aspects involved in the work. The work of R&D allows experts in the different aspects to continue to develop that understanding. Problems occur that require solutions and opportunities for new applications for the technology require its adaptation.

The basic area of research is in microbiology, as anaerobic digestion is a microbiological process (see Chapter 4). Although the process has been well researched over many years, it is still not well understood. The work of R&D in an extension project needs to be more applied, considering the suitability of different feed materials, especially different mixtures of materials, for use in biogas plants. The existence of research facilities allows tests to be made of slurry in plants that have failed, so that the cause of failures can be identified.

A second area of research is in the technical designs of biogas plants and associated equipment. While there are several basic designs of plant that have been proven to be of low cost and very reliable (see Chapter 5), these are not suitable for all conditions. For example, the development of small-scale, low-cost biogas plants for cold climates is still an important area of investigation (Karki *et al.*, 2005).

The work of biogas extension is part of the wider subject of rural development, which has been an important area of research and development. The process of encouraging people to adopt a novel technology that can improve their lives is one that requires careful study (Wargert, 2009). People need to be motivated, not only by the benefits that the technology will bring to their lives, but they also need to feel confident that the technology will continue to be supported (Buysman and Mol, 2013). An important part of the R&D in rural development is finding the most effective ways to provide subsidy and loan finance to people so that they can afford to purchase biogas plants (Awasthi, 2011).

R&D operates at different levels, from laboratory-scale testing to analysis of inter-relationships between different stakeholders involved in a project. Laboratory scale work may involve analysis of the contents of digesters, small-scale tests in the digestibility of different feed materials and mixtures, and the running of small biogas digesters. Digesters may be run as batch or semi-continuous systems, although the results obtained from batch systems may not be directly applicable to semi-continuous systems (Fulford, 1988). A test site, on which several small biogas plants of different designs can be built and tested, is a useful resource. New ideas can be carefully tested before they are made available to customers.

A wider aspect of R&D is the follow up with customers to check the various aspects of the project are working effectively. These visits can allow experienced staff to check the technical performance of the biogas plants and associated equipment to identify areas in which they can be improved and made more reliable. Interviews with customers can also identify weaknesses in the organization of the extension work, including the finance aspects, so that improvements can be made.

A group of people involved in R&D will be keeping track of the work of researchers in other places, as well as documenting their own work and getting it published. Effective international research networks allow ideas to be shared, so people can learn from each other and improve the international understanding of the best ways to build biogas plants, do extension work and provide finance for it.

References

AQP (2004) Health Effects of Wood Smoke, Department of Ecology, Washington State Energy Office, USA. <www.ecy.wa.gov/pubs/92046.pdf> [accessed 18 July 2014].

Ashden (2005a) 'Domestic biogas | The Ashden Awards for Sustainable Energy' [website – case study] Ashden <www.ashden.org/winners/bsp> [accessed 18 July 2014].

Ashden (2005b) 'Kigali Institute of Science, Technology and Management (KIST), Rwanda' [website – case study] Ashden <www.ashden.org/winners/kist05> [accessed 18 July 2014].

Ashden (2006) 'Appropriate Rural Technology Institute: Compact digester for producing biogas from food waste' [website – case study] Ashden Awards <www.ashden.org/winners/arti06> [accessed 18 July 2014].

Ashden (2007) 'Biotech: Management of domestic and municipal waste at source produces biogas for cooking and electricity generation' [website – case study] Ashden <www.ashden.org/winners/biotech> [accessed 18 July 2014].

Awasthi, G.S. (2011) Study on increasing credit access for biogas plants to hills and mountains, Development Vision Nepal (Pvt) Ltd, Baneswor, Kathmandu. <www.snvworld.org/en/download/publications/study_on_increasing_credit_access_for_biogas_plants_to_hills_and_mountains_nepal_2011.pdf> [accessed 18 July 2014].

Bajgain, S. and Shakya, I. (2005) Biogas Nepal - Successful model of public private partnership, SNV (Netherlands Development Organisation) and BSP (Biogas Support Programme). Kathmandu, Nepal <www.snvworld.org/download/publications/bsp_successful_model_of_ppp_nepal_2005.pdf> [accessed 19 July 2014].

Barnett, A., Pyle, L. and Subramanian, S.K. (1978) Biogas Technology in the Third World: A multidisciplinary review, Ottawa: International Development Research Centre.

Buysman, E. and Mol, A.P.J. (2013) 'Market-based biogas sector development in least developed countries - The case of Cambodia', Energy Policy 63: 44–51 <http://dx.doi.org/10.1016/j.enpol.2013.05.071> [accessed 17 July 2014].

de Castro, J., Hossain, Z., Somith, S., Sacklokham, S. and Phan, N. (2008) Asia Biogas Programme Review, SNV Hanoi, Vietnam. <www.snvworld.org/download/publications/asia_biogas_programme_review_2008.pdf> [accessed 22 July 2014].

CDM (2013) CDM Executive Board seventy-fifth meeting, UNFCCC. <http://cdm.unfccc.int/UserManagement/FileStorage/4162R7A9QGHMBCK3PUZYONTF0D5JE8> [accessed 22 July 2014].

CDM (2012) Switch from non-renewable biomass for thermal applications by the user, UNFCCC/CCNUCC CDM Executive Board. <http://cdm.unfccc.int/UserManagement/FileStorage/HSVPWKBG6X7Q8YEFMOT214IA3R0ZDL> [accessed 22 July 2014].

Chen, Y., Yang, G., Sweeney, S. and Feng, Y. (2010) 'Household biogas use in rural China: A study of opportunities and constraints', Renewable and Sustainable Energy Reviews 14: 545–549 <http://dx.doi.org/10.1016/j.rser.2009.07.019> [accessed 17 July 2014].

Devkota, G.P. (2001) Biogas Technology in Nepal: A Sustainable Source of Energy for Rural People, Kathmandu: Mrs. Bindu Devkota, Maipee.

Devkota, G.P. (2007) Renewable Energy Technologies in Nepal: An overview, Kathmandu: Universal Consultancy Services P. Ltd.

Dhakal, N.H. (2008) Financing Domestic Biogas Plants in Nepal, Centre for Empowerment and Development for SNV. Lalitpur, Nepal <www.snvworld.org/en/download/publications/nepal_financing_domestic_biogas_plants_2008.pdf> [accessed 19 July 2014].

Dung, T.V., Hung, H.V. and Hoa, H.T.L. (2009) Biogas User Survey, Vietnam 2007-2008, > [accessed 19 July 2014].

EBA (2011) Biogas, Simply the Best, European Biogas Association. Brussels <http://europeanbiogas.eu/wp-content/uploads/files/2013/10/EBA-brochure-2011.pdf> [accessed 19 July 2014].

EnviTec (2008) EnviTec Forum Biogas Spezial1En Newletter, EnviTec <www.envitecbiogas.com/fileadmin/Newsletter/en/EnviTecForumBiogasSpezial1En.pdf> [accessed 17 July 2014].

Fulford, D. (1988) Running a Biogas Programme: A handbook, London: Practical Action Publications (Intermediate Technology Publications).

Fulford, D., Devkota, G.P. and Afful, K. (2012) Evaluation of Capacity Building in Nepal and Asia Biogas Programme, Hanoi, Vietnam: Kingdom Bioenergy Ltd for SNV. <www.kingdombio.com/Final Report - whole.pdf> [accessed 17 July 2014].

Ghimire, P.C., Chitrakar, J., Malla, B., Ghimire, S., Nepal, K.N. and Misra, S. (1996) Impacts of Biogas on Users, DevPart - Nepal for SNV. Kathmandu, Nepal<www.snvworld.org/en/download/publications/final_report_impacts_of_biogas_on_users_nepal_1996.pdf > [accessed 19 July 2014].

ter Heegde, F. (2005) Domestic Biogas and CDM Financing, Perfect Match or White Elephant, Vietnam: SNV Netherlands Development Organisation,. <http://www2.gtz.de/Dokumente/oe44/ecosan/en-domestic-biogas-cdm-financing-background-2005.pdf> [accessed 19 July 2014].

ter Heegde, F. (2008) Domestic Biogas Projects and Carbon Revenue: A strategy towards sustainability?, SNV Netherlands Development Organisation. The Hague, Netherlands <www.snvworld.org/en/download/publications/domestic_biogas_projects_and_carbon_revenue_2008.pdf> [accessed 17 July 2014].

Idan, J. (2006) 'Presentation on Integrated Sewage and Solid Organic Waste-to-Energy Project', [website – presentation] <www.biogasonline.com/downloads/integrated_waste_to_energy_project.pdf> [accessed 18 July 2014].

Karki, A.B. (1996) A Training Manual for Extension, Kathmandu: CMS for Food and Agriculture Organization (FAO). <www.fao.org/sd/egdirect/egre0021.htm> [accessed17 July 2014].

Karki, A.B., Shrestha, J.N. and Bajgain, S. (2005) Biogas as Renewable Source of Energy: Theory and development, Kathmandu, Nepal: BSP-N <www.snvworld.org/en/download/publications/biogas_as_renewable_energy_theory_and_development_nepal_2005.pdf> [accessed 17 July 2014].

Kossmann, W., Pönitz, U., Habermehl, S., Hörz, T., Krämer, P. and Klingler, B. (1999) Biogas Digest Volume III: Costs and benefits and programme implementation, ISAT and GTZ, Eschborn, Germany. <www.susana.org/lang-en/library?view=ccbktypeitem&type=2&id=1717> [accessed 17 July 2014].

Kotrba, R. (2007) 'German Biogas Firm Expands Global Presence', Biomass Magazine [website - journal] <http://biomassmagazine.com/articles/1360/german-biogas-firm-expands-globalpresence> [accessed 18 July 2014].

Kram, J. (2007) 'A New Day for Biogas: Germany Leads the Way in Europe', Biomass Magazine 1: 24–29 <issuu.com/bbiinternational/docs/bmm-june.07_print> [accessed 18 July 2014].

Lawbuary, J. (2000) 'Biogas Technology in India: More than Gandhi's Dream?', [website - journal] Ganesha <www.ganesha.co.uk/Articles/Biogas Technology in India.htm> [accessed 17 July 2014].

Lohri, C.R., Vögeli, Y., Oppliger, A., Mardini, A.R., Giusti, A. and Zurbrügg, C. (2010) 'Evaluation of Biogas Sanitation Systems in Nepalese Prisons', in IWA-DEWATS, Surabaya, Indonesia: Swiss Federal Institute of Aquatic Science and Technology (Eawag). <www.eawag.ch/forschung/sandec/publikationen/swm/dl/Lohri_2010.pdf> [accessed 17 July 2014].

MyClimate (2012) 'From cow dung to biogas in Karnataka, India', My Climate <www.myclimate.org/fileadmin/myc/klimaschutzprojekte/indien-7149/klimaschutzprojektindien-7149-project-story.pdf> [accessed 19 July 2014].

van Nes, W., Lam, J., ter Heegde, F. and Marree, F. (2009) Building Viable Domestic Biogas Programmes: Success factors in sector development, SNV Netherlands Development Organisation. The Hague, Netherlands <www.snvworld.org/files/publications/snv_building_viable_domestic_biogas_programmes_2009.pdf> [accessed 17 July 2014].

NIIR, B. (2004) Handbook on Biogas and its Applications, Delhi: National Institute of Industrial Research. <www.niir.org/books/book/handbook-on-bio-gas-its-applications/isbn-8186623825/zb,,72,a,5,0,a/index.html> [accessed 18 July 2014].

Puxin (2013) 'Puxin 100 m3 Biogas Digester', [website] Puxin. Shenzhen, Guangdong, China <paterex.com/biog/puxin.pdf> [accessed 19 July 2014].

Roberts, C. (2011) Heating Homes With Human Waste Is Saving Lives And Tigers In Nepal, Fast Company for World Wildlife Fund. Washington DC <www.fastcompany.com/1771107/heating-homes-humanwaste-saving-lives-and-tigers-nepal> [accessed 18 July 2014].

Ross, C.C. and Drake, T.J. (1996) The Handbook of Biogas Utilization, Environment, Health, and Safety Division, Georgia Tech Research Institute Atlanta, Georgia.

Sajidas, A. (2008) 'Biotech: Biogas plant consultancy', [website] Kerala: Biotech Ltd <www.biotech-india.org> [accessed 18 July 2014].

Smith, K.R. (1993) 'The Health Impacts of Biomass and Coal Smoke in Africa', Boiling Point <www.hedon.info/View+Article?itemId=10676> [accessed 18 July 2014].

Stone, J. (2012) Climate Sensitising Nepal's Renewable Energy Sector, Alternative Energy Promotion Centre (AEPC), Lalitpur, Nepal <www.snvworld.org/en/download/publications/carbon_and_climate_unit_climate_sensitising_renewable_energy_sector_nepal_2012.pdf> [accessed 19 July 2014].

Tam, D.M. and Thanh, N.C. (1982) Biogas Technology in Developing Countries?: An overview of perspectives, Bangkok: Environmental Sanitation Information Center, Asian Institute of Technology. <http://nla.gov.au/anbd.bib-an3042819> [accessed 19 July 2014].

Wargert, D. (2009) Biogas in Developing Rural Areas, Sweden: Lund University. <www.davidwargert.net/docs/Biogas.pdf> [accessed 19 July 2014].

Wellinger, A. and Murphy, J. (2013) Biogas Handbook: Science, production and application, D Woodhead Publishing Series in Energy. Cambridge, UK <www.iea-biogas.net/biogas-handbook.html> [accessed 19 July 2014].

Zurbrügg, C. and Volkart, E. (2012) Comparing LCA and CDM Methods - Two ways to calculate greenhouse gas emission reduction due to organic waste treatment, Dübendorf, Switzerland: Swiss Federal Institute of Aquatic Science and Technology (Eawag). <www.eawag.ch/forschung/sandec/publikationen/swm/dl/lca_cdm.pdf> [accessed 19 July 2014]

CHAPTER 4
How biogas works

Abstract

The making of biogas (methane and carbon dioxide) from feed materials is an anaerobic microbiological process, involving a symbiotic population of microbes (bacteria and Archaea*). The process requires the right conditions – pH, temperature, retention time, the chemical and physical nature of the feed, and moisture content. There are several ways in which an anaerobic digester can be run, such as in batch, continuous or semi-continuous modes. Different approaches to the process include using stirred tanks (CSTR), plug flow, upflow sludge blanket (UASB), anaerobic filters, and baffled reactors. There are also different ways to pre-process the feed including chopping and pre-digestion using flooded pre-digesters and leach beds. The quality of the feed material can be monitored using different parameters, such as TS (total solids), VS (volatile solids), COD (chemical oxygen demand), BOD (biological oxygen demand), batch or semi-continuous tests. The process can be mathematically modelled.*

Keywords: biogas, anaerobic digestion, science

The anaerobic process

Although the whole process is very complex and far from fully understood, a simplified view can be presented. All members of a biogas programme, including the administrators, should have this basic understanding.

A biogas plant is a living system composed of large numbers of different of organisms working symbiotically (Finstein, 2010). A cow could be considered to behave as a biogas plant because the animal has a symbiotic relationship with the microorganisms that live in its gut and help it to digest the food it eats. A biogas plant fed primarily on cattle dung can be considered an extension of the gut of a cow, so it should offer a similar environment. This means a biogas plant should be a closed container that is free from light and oxygen and should run at blood heat, a temperature of 35 °C. It should also be gas tight, so that the biogas can be collected.

The population of different organisms in a biogas plant include bacteria and also *Archaea*. Archaea (Madegan *et al.*, 2009) used to be called *archaebacteria* as they are more primitive organisms than bacteria. They are now classified as a separate domain from bacteria as they are seen as very different. The archaea that generate methane descend from organisms that lived on earth before oxygen was generated by plants. They adapted to the presence of oxygen by developing

http://dx.doi.org/10.3362/9781780448497.004

a mechanism that allows them to go into a quiescent state until they encounter suitable anaerobic conditions. Anaerobic digesters must therefore provide the conditions that encourage these organisms to function effectively.

The organisms in an anaerobic digester are opportunistic, taking advantage of conditions that arise in their environment, so a change in conditions can change the population of organisms. If, for example, the material feed to the digester changes, the population of microorganisms that can use the new material will multiply, while those that are less well adapted will go into a quiescent state. An example of this is the addition of a material that contains sulfur to a digester. The organisms that can exploit sulfur will begin to grow and generate hydrogen sulfide, even though they had been quiescent for many months previously. The population will adapt to new conditions, as long as the conditions are changed sufficiently slowly.

The presence of large numbers of different bacteria and archaea in an anaerobic digester is required to break down the long chain molecules in food materials to methane and carbon dioxide. The actual process involves a large number of steps. This is usually simplified to three or four main steps, each of which is composed of many more (Figure 4.1).

The first stage is hydrolysis. This is the break down of materials by the addition of a hydrogen atom (H^+) to one part and a hydroxyl radical (OH^-) to the other. For example, a long chain molecule, such as starch, is broken

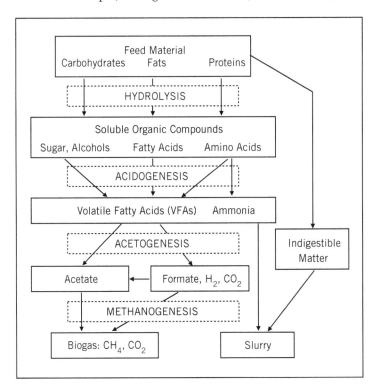

Figure 4.1 Simplified process of anaerobic digestion (adapted from (Husain, 1998))

down into sugars by adding hydrogen or hydroxyl radicals to the oxygen atoms, which act as links between the sugar molecules in the chain. The effect of hydrolysis is to break molecules of carbohydrates, fats, and proteins into shorter molecules, which are more soluble in water.

The bacteria that perform hydrolysis are facultative, i.e. they can live either with or without oxygen. They can actually use oxygen, if it is available, to help the breakdown process. Tests on the start up of a biogas digester that used a pressure gauge to measure the amount of gas in the plant showed that the gas pressure went down over the first day or so after the plant was filled with cattle-dung slurry. The hydrolyzing bacteria were using the oxygen from the air in the plant. These bacteria can also use oxygen from fats and oils to help break down other food materials. The addition of small amounts of glycerine (glycerol) from the manufacture of biodiesel has been shown to accelerate the hydrolysis process (de Rosa and Gambacorta, 1986).

The second stage is acidogenesis; the formation of volatile fatty acids (VFAs). This can be considered as either a single stage or as two stages, with acetogenesis as a second stage. Acidogenesis is the break down of longer chain fatty acids into shorter chain VFAs (C_3 to C_6), such as butyric acid ($CH_3CH_2CH_2COOH$), found in rancid butter, or propionic acid (CH_3CH_2COOH). Acetogenesis refers specifically to the formation of acetic acid (CH_3COOH). Acetic acid is the acid found in vinegar. Fatty acids are generated by further populations of facultative bacteria; those which normally cause food to rot if it is left in the open air. These fatty acids, and especially the volatile fatty acids, plus the aldehydes and ketones that are formed in the process of their decomposition, are the cause of the obnoxious smells that are characteristic of rotting foods and of animal dung.

The final stage is methanogenesis – the generation of methane and carbon dioxide. The methanogenic archaea are obligate anaerobes, which means they cannot function if oxygen is present. They generate biogas from acetic acid (CH_3COOH). Methanogens can also use hydrogen and carbon dioxide and single-carbon compounds such as formate and methanol for methane formation, but acetic acid is the most commonly-used compound (Madegan *et al.*, 2009).

Hydrogen and carbon dioxide are intermediate products formed during the overall process of converting VFA to biogas, and there has been some interest in the evolution of hydrogen as an intermediate product. Several groups have looked at methods to redirect the process to generate hydrogen as the main product instead of methane (Nishio and Nakashimada, 2007). However, the development of this approach is still at the research stage and it is a long way from being an approach that is as established as conventional anaerobic digestion.

Conditions for biogas digestion

Organisms involved in anaerobic digestion only function if the conditions are suitable for them, therefore it is important to define those conditions correctly. Since the methanogens are more sensitive to conditions that are not right for them than other bacteria in the medium, they are usually the first to stop working. They go into a quiescent state to wait until conditions are again suitable. If the methanogens stop, then the other bacterial processes tend to generate excessive VFAs, so the plant goes sour; the pH drops and the slurry begins to smell obnoxious. Due to the fact that acetate and hydrogen are not being used up by methanogens, they will then build up in the medium and result in 'product inhibition' of acetogenesis, in which the reaction is slowed down due to an excess of products relative to reactants (Fox and Pohland, 1994). It has been shown (Hecht and Griehl, 2009) that acetogenesis can also be inhibited by an accumulation of excess longer chain fatty acids. Once a plant begins to become sour, each process seems to become inhibited in turn, so the process collapses as a chain reaction.

It is fairly difficult to recover from a sour condition because the presence of excessive VFAs is inhibiting to methanogenesis. The fastest and sometimes the only solution is to remove the slurry and refill the digestion chamber with new slurry with the right conditions. It may be possible to add the old slurry in small amounts back to the plant once it is running properly again.

pH and alkalinity

The anaerobic process generates fatty acids, but low pH (conditions that are too acidic) can inhibit the process. Anaerobic digestion works most effectively at a pH of close to 7 or slightly above (near neutral). A biogas plant that is working well is self-buffering; the process adjusts its own acidity. Carbon dioxide produced in the process dissolves in water to form bicarbonate ions (HCO_3^-), which causes the solution to be mildly alkaline. The quantity of bicarbonate ions in solution depends on the amount of acids in the slurry. The exchange of carbon dioxide between the gas phase and the liquid phase controls the pH of the slurry.

The pH of a plant can be measured by a pH meter or by using indicator papers that change colour at different pH values. However, pH is an insensitive measure of what is occurring in the plant because of the buffering effect. A low pH value indicates that the plant has already failed, because the buffering mechanism has been overwhelmed by too much acid being generated.

A better measure of the stability of a biogas plant is therefore its alkalinity – the quantity of bicarbonate ions in solution. The proportion of bicarbonate can be measured by titrating the solution against a known acid (Eaton *et al.*, 2005). The amount of acid required to neutralize the bicarbonate is a measure of how much acid the system can tolerate before it begins to fail. There are standard kits available to do this (Hawkes *et al.*, 1994).

An even better measure of plant stability is the ratio of alkalinity from bicarbonate (as measured to pH 5.75) compared to alkalinity from VFA salts (as measured between pH 5.75 and pH 4.3). The bicarbonate alkalinity is what provides the 'useful' buffering capacity for methanogens, but at lower pH (below 5.75) the dissociated VFAs in solution can also accept protons (and lead to a deceptively high value for total alkalinity if only a single-point titration is used). Comparing the ratio of intermediate alkalinity (IA: the amount of acid used between 5.75 and 4.3) to partial alkalinity (PA: acid used to titration point 5.75) gives an indication of the relative amount of VFA in the system. An optimal IA:PA ratio is below 0.5, and a trend of increasing IA:PA ratio indicates an increase of VFA in the system (therefore showing VFAs that are not being converted to biogas) (Ripley *et al.*, 1986).

Temperature

Microorganisms can be designated by the temperature at which they function most effectively. Cryophylic organisms function best at low temperatures; mesophylic organisms function best at medium temperatures; thermophylic organisms function best at higher temperatures. There are certain organisms classed as extremophylic that work best at extreme temperatures. Most of the organisms that take part in anaerobic digestion are either mesophylic, operating most effectively at the temperature in a cow's gut, around 35 °C, or thermophylic, functioning at the working temperature of a compost heap, around 55 °C.

For a mesophylic population the gas production rate increases roughly linearly with temperature between 15 °C and 35 °C ; a 10 °C rise in temperature causes the rate of gas production to approximately double (Madegan *et al.*, 2009). The amount of gas generated per unit mass of feed slightly increases with temperature, but not significantly. The methanogens are also sensitive to sudden changes in temperature. A temperature change of more than 5 °C in one day can cause a digester to become acidic, as the methanogens temporarily stop using the VFAs to generate biogas. In a full-size digester, the thermal mass of a large volume of slurry will reduce the rate at which its temperature can change.

The reaction rate falls slightly as the temperature is increased above 35 °C, but then begins to rise again between 45 °C and 55 °C, the thermophylic region. The rate of gas generation at thermophylic temperatures is roughly twice that at mesophylic temperatures, although the total gas produced per kilogram of feedstock in not necessarily higher. The gas generation rate falls again as the temperature increases to 73 °C, when it ceases (Lee *et al.*, 2008).

Biogas plants have been run successfully at both mesophylic and thermophylic temperatures. However, extra energy is required to keep a digester at the higher temperature. Also thermophylic running is seen as less stable than mesophylic (El-Mashad *et al.*, 2004). The thermophylic methanogens seem to be much more sensitive to changes than the mesophylic ones, so conditions need to be much more strictly maintained.

Anaerobic digestion is a slightly exothermic process; energy is released by the breakdown of the food materials. Some of this energy is utilized by the organisms to grow and reproduce, so very little is available to increase the temperature of the slurry. A biogas plant in a cold environment needs to be well insulated, so that its temperature is not being reduced by heat losses. A way to supply heat to the plant must also be provided.

Some plants use heat exchangers within the body of the digester pit, through which heated water is passed. If there are any problems with these heat exchangers, the digester pit must be emptied, so that they can be accessed (Inglis *et al.*, 2007). A typical problem could be leaks in the pipes causing dilution of the slurry. Fibrous material in the slurry can become entangled around the pipes, and slurry can form a dry crust around the pipes, both of which can reduce the heat transfer rate from the pipes.

Slurry can be pumped from the outlet, passed through an external heat exchanger and put back in the digester. This system can easily be accessed by shutting valves that connect the heat exchanger to the digester and draining the small amount of slurry in the heat exchanger volume. Such a system also allows the slurry to be stirred and organisms recycled within the digester. The simplest approach is to heat the slurry that is being fed to the digester, as long as the digester pit is sufficiently well insulated.

Retention time

The process of anaerobic digestion takes time. Given sufficient time, most organic material can be converted to biogas, but some materials are more easily digested than others. Feed material that is already soluble in water, such as sugars, can be converted to biogas in less than a day, while less digestible organic matter, such as woody lignin, will take several months. Most feed material, such as animal dung, is a mix of different materials, some which can be digested quickly and others which take much longer. At lower temperatures the process takes longer. This means that the retention time, the time for which material is held in the digester for gas to be released, is linked to the temperature at which the digester is run. Lower temperatures mean longer retention times are required.

The pattern of gas production suggests that, at optimum temperatures, much of the gas (at least half of the total) is produced in the first few days that the feed material is placed in the digester. The gas production rate begins to drop with time after the peak is reached. The value of the retention time that is selected is therefore dependent on how much of the available gas is required. A long retention time requires a large total container volume to contain the slurry, so that more gas can be generated. A short retention time requires less slurry, but means that less gas is available from each unit mass of feed material.

For animal dung at mesophylic temperatures (around 35 °C), a typical retention time for a simple digester is 25 days. At lower temperatures, the retention time is longer. For thermophylic digestion, the retention time can be reduced to 10 to 15 days.

Toxins

Any material that is designed to kill bacteria will stop an anaerobic digester functioning. Therefore materials such as antibiotics, cleaning agents and disinfectants should be kept well away from a biogas digester. Chlorine and chlorinated hydrocarbons are very poisonous to anaerobic digestion. The methanogens are the most sensitive of the organisms and will become quiescent quickly if small amounts of these toxins are present. The more resilient bacteria will continue to generate volatile fatty acids, so the plant becomes sour.

This means that the dung from animals fed with antibiotics must not be used in biogas plants. These animals must be kept separate, if animal dung is used as a feed for a plant. If areas that are used for the collection of material to be put in a biogas plant are cleaned with disinfectants, the wash water must be directed away from the plant. Plenty of clean fresh water should then be used to wash the area before it can again be used to collect material to be used in the plant.

There are cleaning agents that are now available that are designed to be less toxic to the environment. These products often use bacteria or enzymes to remove grease and other contamination and these products often have similar functions to the organisms in a biogas plant. Tests need to be done on such cleaning agents to determine whether they are toxic to methanogens in the plant.

Tests on the slurry from an anaerobic digester suggest that it can be toxic to many undesirable bacteria, such salmonella. The volatile fatty acids generated in the process of anaerobic digestion will slowly kill these bacteria (Banks, 2002; Banks *et al.*, 2008). However, the time taken for complete destruction is much reduced at thermophylic temperatures (Smith *et al.*, 2005; Svoboda, 2003).

The physical nature of the feed

Animal dung is considered to be a good material for a biogas plant as it is in the right physical form for the microorganisms to access the food material it contains. The animal has ground up the feed material with its teeth and the material has already been though a digestion process. However, the animal has already used a significant proportion of the energy that was contained in the food material it consumed for its own purposes. The most easily digestible matter has been absorbed by the animal's gut.

Ruminants, such as cattle, process food material very thoroughly. This means that the animal has removed a large amount of energy from its food. The dung can be easily digested in a biogas plant. Many other ruminant animals, such as sheep, goats and deer, produce dung in the form of pellets, covered by a dry outer layer. These pellets do digest well, once the dry outer layer has been broken. The dung from non-ruminant animals, such as pigs, contains more of the available energy in the food, but it is slightly less easily digested. Horses and elephants are much less efficient at removing energy from the food they eat and their dung contains more indigestible fibres.

The use of vegetable matter as a feed potentially provides much more gas per kilogram of material than dung. The demand for processing of food wastes in developing countries is increasing rapidly (Vögeli *et al.*, 2014). In India, for example, the tradition has been to allow cattle to roam cities, and they ate vegetable wastes left at the wayside by householders. Rapid development has meant that the small gardens in which cattle were kept have been sold to developers; disposing of food wastes is becoming a problem. Local authorities do not have a system for collecting such wastes. Disposal of waste from urban vegetable markets is also an issue (Heeb, 2009).

The vegetable matter for the digester must be processed so that the digestible matter is available to the microorganisms. Food materials that are enclosed within a skin, such as unpeeled potato or fruits, cannot be easily digested. Once the skin is broken, the microorganisms can break down the easily digested materials inside it, as well as the skin itself.

Food-processing wastes, such as brewing wastes or vegetable and fruit peelings, have usually been cut or ground up at the food-processing plant as part of the process. These are usually easily digested, although it can be useful if this material can be further chopped or ground. If the food material is in very small pieces, the bacteria can begin to digest it more easily. The hydrolytic bacteria can only digest the surface of a food item, so the more surface area relative to size of the particle, the more potential reaction sites are available. The bacteria take time to penetrate into a piece of food, as the dissolved products of hydrolysis (the products of the first part of the digestion process) need to diffuse away from the surface.

A mixture of different feed materials digests more effectively than feed material from a single source (Alvarez and Liden, 2008; El-Mashad and Zhang, 2010; Molinuevo-Salces *et al.*, 2010). The addition of cattle dung to a mix means that the microbial population is replenished, while the presence of fruit and vegetable matter ensures the microbes have adequate food materials on which to work.

Moisture content

The amount of water in a biogas digester influences the behaviour of the slurry, although it does not directly affect the process of anaerobic digestion, except when the material is very dry or when the slurry becomes very dilute.

Many systems have been run as dry digesters (Kusch and Oechsner, 2008) with a moisture content down to about 50 per cent. A container filled with a dry material, such as hay or straw, and thoroughly wetted with slurry from a working digester, will generate biogas at a reasonable rate. Mixing such dry systems can be difficult, as food materials have a low mobility within the slurry (Ward *et al.*, 2008). Dry systems are often designed without moving parts within the digester; screw conveyors rather than pumps may be used to move the material in and out of the digester, while gas recirculation is often used for internal mixing.

The main reason for the addition of water to an anaerobic digester is to make a slurry that flows through the system and is easily stirred. Water content of 88 per cent or more, or a total solids content of 12 per cent or less, gives a slurry that can be poured from a container or that will flow down a pipe. Slurry that has a higher total solids content becomes a gel, which will adhere to the surface of a pipe and will move much more slowly. Since many simple biogas digester designs use hydraulic pressure to transfer slurry through the system, it is common to add sufficient water to dilute the total solids to below 12 per cent.

Slurry that has a total solids content of less than about 4 per cent will tend to separate into layers. Lighter material will float to the top and heavy material will form sludge at the bottom. The middle layer of liquid has a low total solids content, so organisms that are isolated within this layer have poor access to food materials. This separation into layers was seen as a problem in early digester designs (Fry, 1975), but later operators recognized the value of keeping the total solids content above 4 per cent.

This tendency of different components to naturally separate is used to concentrate the solid materials in many sewage systems. Wastewater from a drainage system usually includes washing water as well as sewage, so has a total solid content of less than 1 per cent. Placing this mix in a large tank allows the solid material to settle to the bottom as sludge. The liquid layer has a low concentration of food materials, so can be easily cleaned by aerobic techniques. The sludge is removed and processed in anaerobic digestion systems.

Carbon to nitrogen (C:N) ratio

The organisms in an anaerobic digester use the carbon, hydrogen and oxygen in the feed materials both as a source of energy and as a supply of biomass that help them grow and reproduce. As well as these main building blocks, living organisms need small amounts of other elements, especially nitrogen to enable them to grow. Most biomass contains nitrogen, but the quantity can vary over a wide range. Most mammals produce nitrogen-rich urine, so the dung has a low nitrogen content. If a cow, for example, is fed with a low nitrogen diet, its dung can have a high carbon to nitrogen (C:N) ratio, unless the urine is collected along with the dung.

Nitrogen can also be toxic to the anaerobic process. Excessive amounts of ammonia, the anaerobic breakdown product of nitrogen, can slow the production of biogas. Although the ammonia provides some buffering, this can disguise high concentrations of VFA in the digester and can result in souring of the plant. Birds, such as chickens, do not secrete urine so their dung contains a much higher proportion of nitrogen. Biogas plants run with a high nitrogen content can become unstable (Braun *et al.* 1981).

The optimum C:N ratio is between 25:1 and 30:1 (Hills, 1979; Ward *et al.*, 2008), although biogas plants seem to operate over a much wider range. Values away from this optimum can cause biogas production to be reduced. It is recommended that a C:N ratio below 10 is not used for single-stage digestion.

The form of the carbon and nitrogen is also important. Both must be in a soluble form that is freely available to the microorganisms. However, since the hydrolyzing bacteria break down insoluble feed material to soluble chemicals, it is the soluble C:N ratio in the body of the plant that is the important parameter to be measured. Carbon that is in the form of lignin is not easily released by the hydrolysing bacteria, so cannot offset excess nitrogen in the plant. Nitrogen in the form of ammonia seems to have a stronger influence than nitrogen in other forms (Braun and Huber, 1981; Chen and Cheng, 2008; El-Mashad *et al.*, 2004).

If a feedstock has a C:N ratio that is away from the optimum value, the simplest solution is to find an easily available material that can add carbon or nitrogen (Yen, 2001). Algae can be used to add nitrogen, while cut grass or shredded paper can be used to add carbon.

Mixing

Anaerobic microorganisms have limited mobility and therefore the slurry in an anaerobic digester should be mixed so that the different bacteria and archaea are more likely to come in contact with the particular food materials that they can use. Mixing has a greater effect on high-rate digesters, fed with high concentrations of food material, than on simpler digesters. Mixing should be gentle or periodic rather than constant, because these organisms take time to use food materials. Vigorous mixing has a negative effect (Kaparaju *et al.*, 2008).

A second reason for mixing slurry is that a floating scum layer often forms on top of the slurry. This layer is composed of oil, animal hair and fibrous materials in the feed. If left for a time, this scum can build up and dry out, preventing the gas that is generated in the slurry being able to escape from the top. The gas often finds its way out though other pathways in the system, and escapes into the atmosphere. Gentle stirring, especially of the surface of the slurry, prevents the build up of the scum layer and allows gas to collect in the correct place.

Mixing has been done in several ways depending on the digester design. The simple approach is the use of a paddle or a propeller within the tank. This approach brings technical issues, such as the need to seal the rotating shaft against liquid or gas leakage. Any problems with the parts of the mechanism within the digester means that the slurry must be removed before repairs can be made. Another approach is to compress some of the gas that is produced in the digestion process and bubble it back into the bottom of the digester.

Some of the slurry can be removed from the outlet of the digester and pumped back into it. This approach has the advantage that stirring, heating and recycling of organisms can all be done using a single external system. Any problems, such as the failure of a pump, can be solved without needing to remove the slurry from the digester pit.

The feed material needs to be well mixed into a slurry before it is added to the digester. The hydrolytic bacteria cannot access the insides of large lumps of denser feed material. The slurry needs to be in the form of a smooth paste or soup.

Types of biogas digester

While biogas digestion follows a defined process, there are different ways in which a plant can be operated to take advantage of this process (IEA, 2001). Plants can be run under batch, semi-continuous or continuous conditions. The microbes can be left in free suspension in the slurry or they can be given supports on which they can grow.

Batch digestion

The simplest way to run a digester is to place a mix of microbes and feed slurry in a container and to collect the gas as it is given off. The starter batch of microbes is usually taken from a digester that is already working with a similar feed material to ensure they are already adapted. This starter can be between 5 per cent and 30 per cent by volume of the total volume of the container (Maramba, 1978).

There is a lag time between closing the digester and the commencement of gas generation. This can be between 1 and 14 days. The lag time is least with optimum temperatures (depending whether the system is run at mesophylic (35 °C) or thermophylic (55 °C) temperatures) and a high proportion of starter. The hydrolytic microbes need to remove oxygen from the system and start the breakdown process, thus allowing the methanogens to start working to generate biogas.

The gas production rises to a peak after several days, but then drops off steadily, as the feed materials are consumed. The drop in gas production follows an exponential decay model, which suggests that it is asymptotic, i.e. it approaches zero over time, but never reaches it. Tests on batch digesters that have been left for over a year suggest that biogas is still being produced, although at a very low rate.

Gas from a landfill site can be considered to be batch digestion. Organic matter that is included in the waste material placed into a hole in the ground will degrade anaerobically, once the landfill cell in which it is placed is filled and capped. Landfill gas can be economically extracted and used typically for 20 years afterwards. Methane will still be produced for many more years after this time, but the amount will have reduced to the point when it is not worth extracting.

One benefit of batch digestion is that material that contains indigestible lignin can be loaded into the digester. The digestible content will be broken down into biogas while the digester is being run. Once the gas production rate has dropped to a certain value, the remaining material can be removed and composted aerobically. The remaining material is nearly free of smell, because the digestion process has removed most of the VFAs. As the starter material from a working digester contains microbes that can operate in aerobic conditions, the composting process will occur much more rapidly than if that same material were composted directly without having been through an anaerobic digester.

Another benefit is that the feed material can have a high total solids content, enabling it to be loaded into and out of the digester mechanically. Effective mixing of the solid material with the liquor containing the microbes can be done by removing the leachate from the bottom and spraying it back over the top of the solid fill (Huy, 2008).

The disadvantage, of course, is the variable gas production rate. This can be overcome by running a set of batch digesters sequentially. One digester can be emptied and refilled at a point in time, while the others are at different stages in their cycle. A set of 31 batch digesters will have 30 tanks producing gas, while one of them is emptied and filled each day (Maramba, 1978). A 30-day retention time is adequate in a tropical situation, when the ambient temperature is usually above 25 °C or more and the typical retention time is 30 days.

Continuous digestion

The problem of variable gas production can also be overcome by feeding a single plant continuously. Since all the organisms in a digester need access to the particular feed materials that they use to function and grow, the processes can work more effectively if they are happening simultaneously. A steady flow of feed material allows this to occur; each organism has access to the breakdown products generated by the previous organism in the chain.

A strict definition of continuous feeding is that the feed material is pumped into the digester at a steady slow rate. Laboratory-scale digesters have been designed to run in this way, but this approach adds complications to a full-scale digester. An extra tank is required from which the feed material is supplied and a pump must be used.

Biogas plants run in a continuous mode when feed is added at regular intervals. Most domestic-scale systems are fed once a day and produce biogas at a fairly steady rate. The feeding interval can be increased to every two days. There may be an increase in gas production in the hours after the plant is fed, with a corresponding drop later on, but these variations are usually only a small percent of the average gas production. Plants that are fed once a day are sometimes classed as semi-continuous.

The feed rate of a continuous plant is an important parameter because the rate of gas production is related to the rate at which feed materials are added to the plant. However, there is a limit to the rate at which feed can be added, above which the gas production begins to drop or at which the plant begins to fail and go sour. This occurs when the acidogenic bacteria are generating VFAs faster than the methanogens can use them (Zoetemeyer *et al.*, 1982).

Semi-batch operation

Some biogas plants in China are fed with a mixture of animal dung, human sewage and vegetable matter. These plants generate gas continuously, but there is a build up of lignin and other indigestible matter in the digester pit.

These plants are completely emptied every six or twelve months to remove this material and are restarted again. The slurry that is removed is usually placed on a compost heap and used as fertilizer in the fields.

There is a period when gas is not available, while the plant is being emptied, refilled and restarted. Some of the liquor from the previous load can be retained and used as a starter to reduce the lag time in the starting of the digester.

Continuous stirred-tank reactor

The simplest continuous anaerobic digester relies on the microbes staying in suspension in the slurry. This will occur if the total solid content of the slurry is more than about 4 per cent or the slurry is being stirred. The term for this type of reactor is continuous stirred-tank reactor (CSTR). The addition of new slurry at regular intervals provides a simple stirring effect, even if a mechanical stirring system is not provided.

The main benefit of a CSTR is its simplicity; all it needs is an empty tank to contain the slurry and a system for collecting the gas that is generated. A method for regularly feeding material into the plant is required, as well as a slurry outlet, so spent slurry can be removed. By placing the inlet pipe opening at a level that is higher than the outlet from the main digester, old slurry will flow out of the outlet pipe, as new slurry is poured in the inlet, provided the slurry has a total solids content of less than 12 per cent.

One weakness of a CSTR is that new slurry is mixed with the complete volume of old slurry. This implies that a small proportion of new slurry will flow from the outlet before it has been fully digested. If the inlet and outlet pipes are places on opposite sides of the tank and the stirring is gentle, the new material will take some time to migrate across the volume. This does not exclude the possibility of short circuit, but it does reduce its significance.

Plug-flow digester

One approach to reducing the short-circuit effect is to increase the distance between the inlet and outlet pipes by making the digester vessel long and narrow (Lastella *et al.*, 2002). This approach is called plug-flow because when a volume of slurry is fed to the plant, it moves along the vessel as a plug. This plug goes through each stage of the digestion process as it moves along. Slurry from the outlet of the plant, the effluent, must be mixed into the slurry entering the plant, so the microbes can be recycled. Early tests (Liu, 1998) on plug-flow plants suggested that they were more efficient at generating biogas. However, this is not always valid, as conditions can be adjusted in both systems that allow the CSTR system to be more efficient (Li *et al.*, 2007).

Wastewater digesters

Slurries with low total solids content (less than 4 per cent) tend to separate into different layers if placed in a CSTR (Fry, 1975). Therefore a CSTR is less effective and different approaches are required for dilute slurries, such as the wash water from food processing. Various designs of digester have been developed to handle dilute slurries. The aim is to provide a large surface area on which the microorganisms can grow, although they will do this for themselves, forming a sludge blanket, if the conditions are right.

The addition of additional items within a digester (such as those discussed below) does increase the cost of a system. It also increases the risk of blockage. These more elaborate systems are required for the more dilute feed materials, but their use can be counter-productive for feedstocks with a total solids content of more than 4 per cent.

Upflow anaerobic sludge blanket digester

The upflow anaerobic sludge blanket (UASB) system uses the tendency of dilute slurry to separate; the solid material forms a layer (a blanket) at the base of the digester vessel (Seghezzo *et al.*, 1998). The slurry is gently pumped upwards through the sludge blanket, which traps the food materials in the water and digests them to biogas.

The microbes in the sludge blanket form themselves into granules of between 0.5 mm and 2.0 mm in diameter. These granules are composed of fibrous microorganisms that clump together. They are heavier than water and sink. They are porous, so they filter the water as it flows through the blanket and allow methane and carbon dioxide to be released.

An expanded granular sludge bed (EGSB) is a variant of the UASB (Austermann-Haun, 2008) in which the flow rate of the water is higher, so the granules are dragged upwards by the flow, making the blanket much deeper. This approach is valuable if the waste content of the water is very low. The wastestream can be recycled through the sludge bed by pumping some of the effluent stream back to the inlet, so increasing the chance of removal of polluting materials.

A further development of the UASB is the induced bed reactor (IBR), which includes a three-phase separator, which directs solids, liquids and gases to be passed in pre-determined directions (Dustin, 2010; Finstein, 2010). The IBR is capable of treating wastewaters with a higher concentration of solids than a normal USBR can process.

Anaerobic filters

A sludge blanket acts as a natural filter, but the granules take time to form. Inert support media can be placed inside a digester tank to which the microorganisms can become attached. Many different types of support media have been tested and shown to function. Early designs (Brower and Barford, 1997) used ceramic materials or sand, which remained at the bottom of the digester. Plastic materials appear to work in the same way (Saravanan and Sreekrishnan, 2006) and can float within the slurry. Activated carbon has often been used, as it also floats and has a large surface area on which the bacteria can find attachment sites.

The support media can be in the form of pieces that sit within the slurry in the digester or as curtains or walls that are fixed to the sides or roof. Curtains can also be attached to stirrer bars that can be moved within the digester vessel increasing the chance of contact between the attached microorganisms and the food materials.

Baffled reactors

The support media can be in the form of fixed walls or baffles that force the flow of wastewater to follow an extended path through the digester (Barber and Stuckey, 1999). A baffled reactor can also be considered as a form of plug-flow digester. The baffle walls can be plain or they can be covered with support media to further increase the surface area.

Preparation of the feed material

The feed material needs to be in the right state for the organisms to work on it. It needs to be free of toxins and chopped up into small pieces; it may need to have water added; it needs to have the right C:N ratio; and it needs to be mixed to be a homogeneous paste or soup.

Separation of fibrous materials

Biomass contains mainly carbon, hydrogen and oxygen, with some nitrogen and other elements in much smaller quantities. However, the carbon, hydrogen and oxygen come as a very wide range of different molecules, some of which are easier to degrade than others. The main constituents of plants are hemicellulose, cellulose and lignin, which are all polysaccharides. However, these three materials vary in their stability and the ease with which they can be broken down. Hemicellulose is easily degraded, while cellulose is more difficult. While lignin can be degraded biologically, it takes a long time and requires fairly specialist enzymes. Often cellulose and hemicellulose are bound within lignin complexes, so cannot easily be reached by bacteria. So, materials with high lignin content, such as wood, will not be digested in an anaerobic digester.

In mixed feeds, materials containing lignin and other indigestible matter, such as the protein keratin, will be left as fibres in the slurry. Examples of such fibres are lengths of straw, animal hair and the woody ribs of leaves. These can become entangled and cause blockages in pipes and pumps. They can become wrapped around the shafts of stirrers and pumps and cause them to jam. They often float to the surface of the slurry and form a scum that can dry out to become a solid crust.

There are two main ways to deal with such materials in a digester. Either they can be chopped into very small pieces, so they remain in suspension, or they can be removed from the slurry before it is placed into the digester. If they are finely chopped, hydrolytic bacteria do have a chance to begin to break down these materials. However, the extra energy gained from the gas these materials may produce is often outweighed by the extra energy required to physically break these tough materials into small pieces.

One way to remove fibrous materials from the slurry before it enters the main digester is to use their tendency to form a scum layer on the surface. The feed material can be placed in a container where it is mixed with water to form slurry. The fibrous materials that float to the surface can be skimmed off before the slurry is allowed to flow into the main digester.

The use of pre-digestion

An extension of this approach is to use a pre-digester. The hydrolytic step of the process can be separated from the later stages (Wang *et al.*, 1999; Ward *et al.*, 2008). Liquor extracted from the slurry in the outlet of the plant is used to mix with the new feed material, so that the organisms are recycled. This also reduces the requirement for fresh water to prepare the feed slurry. The feed material and the effluent liquor should be mixed to a homogeneous soup. The slurry can then be left to settle, to allow the fibres to float to the surface so they can be removed. Separating the hydrolysis stage from the later stages allows feedstocks to be processed at a greater rate (Cohen *et al.*, 1982, 1980).

This stage can be accelerated by heating it to thermophylic temperatures (Demirer and Othman, 2008). The mixing container needs to be well insulated and it can be heated using solar water-heaters (Demirer and Othman, 2008; El-Mashad *et al.*, 2004; Lee *et al.*, 2008) or by using the excess heat from an engine. The hydrolytic bacteria work most effectively at 55 °C (Lee *et al.*, 2008) and will aggressively break down insoluble material. The addition of materials that offer an easily available source of oxygen (such as glycerol from biodiesel manufacture) can increase the rate of break down.

A further step is to add air or oxygen to the process (Martinez-Garcia *et al.*, 2007). Since the hydrolytic bacteria are facultative, they can use oxygen to accelerate the breakdown process. A commercial system using a thermophylic aerobic pre-digester was developed at the Bhabha Atomic Research Centre (BARC) in Mumbai, India. The system was designed for domestic food wastes and is called the Nisargruna process (Shah, 2006). The aerobic first stage is heated using solar water-heaters. A thermophylic aerobic first stage allows food wastes to be efficiently processed. There is a danger that the aerobic bacteria can convert easily digestible materials directly into carbon dioxide and water, reducing the amount of methane generated in the second stage digester. Using an aerobic first stage with animal dung as the main feed could be counter-productive.

The Bioplex process (Reynell, 2008) is a thermophylic first stage batch system which was developed to separate cattle dung from straw bedding, but works for any feed materials with a high fibrous content. The portagester is a trailer which is half filled with biomass mixed with fibrous matter to which digester effluent is added and then heated to 55 °C. It is then allowed to settle so the fibres float to the surface and inert solids, such as soil and stones, sink to the bottom. The liquor is then drained into the second stage digester, leaving the straw and solids to be tipped onto a compost heap.

Leach-bed digestion

Flooded pre-digesters, such as the Nisargruna and Bioplex systems, are run in batch mode. A leach-bed digester is designed so that the effluent liquid from the main digester is sprayed over the biomass in the pre-digester container. This liquor contains microbes that start the digestion process, hydrolyzing the food materials and converting them into soluble intermediates, such as long chain fatty acids (Browne et al., 2013). The liquor is then transferred to the main digester, where it is used by methanogen microbes to generate biogas.

The second-stage reactor is producing biogas from dissolved solids, which suggests that it can be of a high-rate design, such as a UASB, or similar, or a filter bed or baffled design (Cysneiros et al., 2011). In a CSTR, it is suggested that the microbes use fibres in the slurry as support media. If the fibres are retained in the leach bed, they need to be replaced by alternatives, such as the granules that form in a UASB, or the media that form part of the design of a filter-bed reactor.

The way the liquor flows through the leach bed is important (Uke and Stentiford, 2013). A solidly packed bed of biomass material causes low flow rates and the formation of localized channels. This means that a proportion of the biomass is not wetted by the liquor, so does not come in contact with the microbes. The addition of fibrous biomass material, such as straw, can allow looser packing and more uniform flow of the liquor through the bed.

Once the solid material in the bed has been thoroughly washed and the easily digestible material dissolved away, the remaining fibrous biomass can be drained and then dug out of the leach bed container. This material is in the form of friable compost. This compost does not have an obnoxious odour, as the main causes of the odours, the VFAs, have been washed into the main digester. This compost can be placed in a heap to 'mature' (to allow remaining volatile chemicals to evaporate) before it is used as compost.

Parameters to measure anaerobic digestion

When running a biogas digester, measurements need to be taken to ensure that the conditions are correct for digestion to take place efficiently. Parameters such as temperature, pH and alkalinity, and moisture content are fairly easy to measure. Measuring the C:N ratio requires a laboratory that can analyse free carbon and free nitrogen. The digestibility of the feed material is the parameter that is of most interest in predicting the amount of biogas that can be generated from that material. This is a parameter that is more difficult to measure.

Total solids measurement

Since the moisture content of slurry is usually high (80 per cent to 96 per cent), it is more common to quote the total solids content. A sample can be removed from the slurry, weighed and dried in an oven at 105 °C for 24 hours. The weight of the dried sample gives the weight of the solid material in the original sample. The value is usually quoted as % TS or sometimes % DM (dry matter).

The dry-matter content of slurry will include inert materials, such as clay and sand, and indigestible organic materials, such as lignin, as well as digestible materials. Any formic acid in the slurry will also be evaporated because it has a boiling point very close to that of water. The other VFAs have boiling points that are higher than 105 °C, so will remain in the solid sample.

Volatile solids measurement

The simplest way to identify the amount of inert material in slurry is to pyrolize it. The sample that is left after drying is placed in a furnace and heated to 550 °C for at least 1 hour (often for much longer). Volatile matter in the sample is thermally converted into gases and vapours, which need to be vented from the furnace. These gases and vapours have a noxious smell, so must be removed from a closed laboratory via a fan extraction system. The weight of the material remaining is a measure of the non-volatile matter in the sample. The difference between the weight of the original solid matter and the non-volatile matter is the volatile solid content, usually quoted as % VS.

This measurement is fairly quick and easy to make, but has several inaccuracies. Indigestible materials, such as lignin, will decompose when pyrolized at 550 °C. Where indigestible biomass is present a VS measurement gives a high measure of the digestibility of the biomass. Where the biomass, such as starch or sugar, is easily digestible, the opposite problem occurs. Most biomass contains a larger proportion of carbon than that in the gases and vapours that are produced in pyrolysis. Some of the carbon is left in the residue as char, giving too low a measure of the digestible matter in the sample.

Chemical oxygen demand (COD)

Biomass material that pollutes water will be decomposed by bacteria in the water. Since the first stage of decomposition is hydrolysis, the bacteria use the most easily accessible source of oxygen they can find, which is the oxygen dissolved in the water. Such pollution is harmful to the fish and other creatures that rely on that oxygen. Chemical oxygen demand is designed to replicate that process chemically, to determine how much oxygen is required to oxidize the pollution in the water. The same technique can be used to determine the amount of digestible matter in a sample of feed material in slurry in a biogas plant (Comino *et al.*, 2012).

The usual oxidant is potassium dichromate in sulfuric acid. The mixture needs to be heated for several hours at a temperature above boiling (usually under reflux) to ensure all the material is oxidized. The mix is then titrated against a reducing agent, such as ferrous ammonium sulfate, to determine the remaining available oxygen.

Although the standard procedure needs careful safety precautions and is labour and time consuming, there are several automated kits available that make the measurements less laborious and much easier to do. The use of smaller quantities of self-contained reagents also makes the procedure much safer to do.

Biological oxygen demand (BOD)

A chemical oxygen demand will include biomass that can be oxidized chemically, such as lignin, so oxidizing the sample biologically should give a more accurate reading. A standard aerobic culture is added to the sample, which is then incubated at 20 °C for five days in the dark. The result is often classed as BOD_5 (Comino *et al.*, 2012).

There are two ways to measure the BOD: the manometric and dilution methods. In the first approach, the amount of carbon dioxide that is generated is measured. This is often done by absorbing it in lithium or ammonium hydroxide, which will reduce the pressure in a sealed container, as oxygen is consumed. In the second method an oxygen meter is used to measure the dissolved oxygen before and after incubation. However, the aerobic culture will also use oxygen, so a blank sample is also required which is exactly the same as in the first sealed container, but does not contain any of the sample. The BOD of the sample is then the difference between the container containing the sample and the container with the culture on its own.

Again, there are several commercial kits that are available to measure BOD. Standard containers, cultures and measuring apparatus mean that the results are reproducible and that different feed materials can be properly compared.

Batch tests

BOD measurement use aerobic microorganisms that are able to degrade more materials than anaerobic microbes. A BOD measurement may not be the most accurate estimate of the digestibility of a feedstock. A better way to measure digestibility would be to add the feedstock to an anaerobic culture and measure the amount of biogas that is generated.

A culture from a working biodigester is added to a sample of the feedstock in a flask, which is sealed with a bung that holds a pipe that leads to a gas collecting system. The flask is placed in a water bath that keeps the temperature at 35 °C and the amount of gas generated is measured over time. The system can be kept running for 30 days, which gives a good accuracy for the gas production plot that allows various constants to be derived.

The simplest gas collecting system is a second flask filled with water, with a bung in the top that carries two pipes. The first pipe allows the gas to enter the top of the flask. The second pipe extends to the bottom of the flask and transfers displaced water to a third flask. The volume of displaced water is measured each day and replaced in the second flask.

Rather than using pure water for measuring the gas, which can absorb carbon dioxide, it is better to use acidulated brine. Common salt (sodium chloride) is dissolved in water and hydrochloric acid added to give a pH below 4.

Gas meters can be purchased to measure the small amounts of gas generated from such system from specialist suppliers, but they can be difficult and expensive to obtain.

Laboratory-scale plants

Tests that compare batch digestion with semi-continuous systems on the laboratory scale suggest that results from one system do not apply to the other (Chowdhury and Fulford, 1992). The only way to reliably test the gas production from a biogas plant is to run a laboratory-scale trail of the system. Since anaerobic digestion occurs at the microbial level, the scale of a digestion system appears to have little influence on its performance. The results from a laboratory-scale system using digester tanks with a volume of only 1 litre can be used to predict the behaviour of a plant with a volume of 100,000 m³. The main difference between large and small scales is related to heat transfer and heat capacity. Laboratory plants need to be kept at the right temperature using a water bath or a temperature controlled enclosure.

Modelling of anaerobic digestion

There is a range of ways to model the process of anaerobic digestion of varying complexities.

Simple models

The simplest approach uses a single parameter, usually defined as biogas potential. Many publications (Biosantech *et al.*, 2013) quote the volume of biogas that can be generated from 1 kg of different feed materials, such as cattle dung or different crops, such as fodder maize.

This very simple model ignores the rate at which biogas is generated. A slightly more complex model is the first order rate model (Banks and Heaven, 2013; Fulford, 1988), which has two parameters, biogas potential, and rate constant, which is a measure of how fast a material generates biogas (see Appendix 1). Calculating two parameters demands much more data from experiments than calculating one parameter.

The 'classic' model of anaerobic digestion was derived by Chen and Hashimoto (Chen and Hashimoto, 1978) from the Contois model of bacterial growth. This includes the ultimate methane yield, a parameter linked to the biogas potential, a dimensionless kinetic parameter, but also a range of other parameters related to the growth of bacteria (Hashimoto *et al.*, 1981). A more empirical model was developed by Hill (Hill, 1991), which included a range of other parameters, including a stress index. The models were seen as relatively simple, but limited in application, as they were developed on the basis that animal dung was seen as the main feed material (Karim *et al.*, 2007). Given the range of parameters that need to be defined, the usual way in which these models are used is to assume that the values of many of the parameters are the same as that measured when the model was first developed. When different experimental conditions are used, the models give less accurate results (Husain, 1998) compared to those measured. However, they can be used to provide a rough idea of the amount of gas that can be generated from a biogas plant (Harris, 2012).

More complex models

With the availability of modern computing power and modelling software, the models for digesters have become more complex. The International Water Association developed a very detailed model of each stage of the digestion process (ADM1) (Batstone *et al.*, 2002), using equations describing both the microbiological processes and the physical processes related to them, such as mass balances and gas mass transfer (Wellinger and Murphy, 2013). Such a model allows the rate-limiting step to be determined, which controls the rate at which biomass can be consumed and biogas generated (Pavlostathis, 2011). The rate-limiting step is usually assumed to be methanogenesis; this is the process that collapses when the system is placed under stress, i.e. when the feed rate is steadily increased. A biogas plant goes sour, i.e. has a decreasing pH, when the methanogens cannot consume the VFAs fast enough to keep up with the acid-forming bacteria.

The ADM1 model has been further developed to apply to a wider range of feed materials and operating conditions (S García-Gen *et al.*, 2013; Santiago García-Gen *et al.*, 2013). These models offer designers and operators an idea of how an anaerobic system should work. However, such models also have limitations, for example, they cannot predict how a system will operate under novel conditions and with new feed materials (Lauwers *et al.*, 2013).

Thermodynamic models

Another approach to the process of anaerobic digestion is to see it as a black box and consider the input and output from it. This was an approach suggested by A.M. Buswell in 1932 (Buswell, 1932). Energy and mass balances can be used to analyze the process. Biomass is primarily formed of carbon, hydrogen and oxygen (C, H, and O), which are converted into methane (CH_4) and carbon dioxide (CO_2) (see Appendix 1). If the original amounts of carbon, hydrogen and oxygen in the feed material are known, or can be estimated, the Buswell equation can be used to calculate the relative proportions of methane and carbon dioxide generated. It also allows the energy generated by the process to be calculated (Banks, 2011). The analysis shows that the heat energy generated from the anaerobic process is only about 5 per cent of that of an aerobic process in which food material is converted to carbon dioxide and water.

Comparing this analysis with actual results from experiments demonstrates that most biogas systems are well below 100 per cent efficient (Mata-Alvarez *et al.*, 2000), especially when fed with biomass other than animal dung. Certain systems, when used with fruit and vegetable processing residues, are only 58 per cent effective (Bouallagui *et al.*, 2005). Some two-stage digesters have efficiencies that are much higher, up to 96 per cent of volatile solids converted to biogas, if the second stage is a high-rate digester.

References

Alvarez, R. and Liden, G. (2008) 'Semi-continuous co-digestion of solid slaughterhouse waste, manure, and fruit and vegetable waste', *Renewable Energy* 33: 726–734 <http://dx.doi.org/10.1016/j.renene.2007.05.001> [accessed 22 July 2014].

Austermann-Haun, U. (2008) 'Anaerobic treatment - A review', *Gas Wasserfach Wasser Abwasser* 149: S6–S11.

Banks, C.J. (2011) 'Anaerobic digestion and energy' [Presentation] University of Southampton <http://www.valorgas.soton.ac.uk/Pub_docs/JyU%20 SS%202011/CB%204.pdf> [accessed 22 July 2014].

Banks, C.J. and Stringfellow, A (2002) 'Biodigestion of kitchen wastes', in: *Proceedings of the Chartered Institute of Waste Management 2002 Conference Torbay, UK.* <http://eprints.soton.ac.uk/53869/> [accessed 22 July 2014].

Banks, C.J., Chesshire, M. and Stringfellow, A. (2008) 'A pilot-scale comparison of mesophilic and thermophilic digestion of source segregated domestic food waste', *Water Science and Technology* 58: 1475–1481 <http://eprints. soton.ac.uk/52628/> [accessed 17 July 2014].

Barber, W.P. and Stuckey, D.C. (1999) 'The use of the anaerobic baffled reactor (ABR) for wastewater treatment: a review', *Water Research* 33: 1559–1578 <http://dx.doi.org/10.1016/S0043-1354(98)00371-6> [accessed 22 July 2014].

Batstone, D.J., Keller, J., Angelidaki, I., Kalyuzhnyi, S.V., Pavlostathis, S.G., ... Vavilin, V.A. (2002) 'The IWA Anaerobic Digestion Model No 1 (ADM1)', *Water Science and Technology* 45: 65–73 <www.iwaponline.com/wst/04510/ wst045100065.htm> [accessed 22 July 2014].

Bouallagui, H., Touhami, Y., Ben Cheikh, R. and Hamdi, M. (2005) 'Bioreactor performance in anaerobic digestion of fruit and vegetable wastes', *Process Biochemistry* 40: 989–995 <http://dx.doi.org/10.1016/j.procbio.2004.03.007> [accessed 17 July 2014].

Braun, R., Huber, P. and Meyrath, J. (1981) 'Ammonia toxicity in liquid piggery manure digestion', *Biotechnology Letters* 3: 159–164 <http://link.springer.com/article/10.1007/BF00239655> [accessed 22 July 2014].

Brower, J.B. and Barford, C.C. (1997) 'Biological fixed-film systems', *Water Environment Research* 69: 487–500 <www.jstor.org/stable/25044903> [accessed 22 July 2014].

Browne, J.D., Allen, E. and Murphy, J.D. (2013) 'Improving hydrolysis of food waste in a leach bed reactor', *Waste Management* 33: 2470–2477 <http://dx.doi.org/10.1016/j.wasman.2013.06.025> [accessed 17 July 2014].

Buswell, A. (1932) *Anaerobic Fermentations*, Illinois: Department of Registration and Education, State of Illinois, USA <webh2o.sws.uiuc.edu/pubdoc/B/ISWSB-32.pdf> [accessed 22 July 2014].

Chen, Y., Cheng, J.J. and Creamer, K.S. (2008) 'Inhibition of anaerobic digestion process: A review', *Bioresource Technology* 99: 4044–4064 <http://dx.doi.org/10.1016/j.biortech.2007.01.057> [accessed 17 July 2014].

Chen, Y.R. and Hashimoto, A.G. (1978) 'Kinetics of methane fermentation', *Biotechnology and Bioengineering Symposium* 8: 269–282.

Chowdhury, R.B.S. and Fulford, D.J. (1992) 'Batch and semi-continuous anaerobic digestion systems', *Renewable Energy* 2: 391–400 <http://dx.doi.org/10.1016/0960-1481(92)90072-B> [accessed 17 July 2014].

Cohen, A., Breure, A.M., van Andel, J.G. and van Deursen, A. (1980) 'Influence of phase separation on the anaerobic digestion of glucose--I maximum COD-turnover rate during continuous operation', *Water Research* 14: 1439–1448 <1354(80)90009-3> [accessed 17 July 2014].

Cohen, A., Breure, A.M., Van Andel, J.G. and Van Deursen, A. (1982) 'Influence of phase separation on the anaerobic digestion of glucose. II. Stability, and kinetic responses to shock loadings', *Water Research* 16: 449–455 <http://dx.doi.org/10.1016/0043-1354(82)90170-1> [accessed 17 July 2014].

Comino, E., Riggio, V.A. and Rosso, M. (2012) 'Biogas production by anaerobic co-digestion of cattle slurry and cheese whey', *Bioresource Technology* 114: 46–53 <http://dx.doi.org/10.1016/j.biortech.2012.02.090> [accessed 17 July 2014].

Cysneiros, D., Banks, C.J., Heaven, S. and Karatzas, K.A.G. (2011) 'The role of phase separation and feed cycle length in leach beds coupled to methanogenic reactors for digestion of a solid substrate (Part 2): Hydrolysis, acidification and methanogenesis in a two-phase system', *Bioresource Technology* 102: 7393–7400 <http://dx.doi.org/10.1016/j.biortech.2011.05.042> [accessed 17 July 2014].

Demirer, G.N. and Othman, M. (2008) 'Two-phase thermophilic acidification and mesophilic methanogenesis anaerobic digestion of waste-activated sludge', *Environmental Engineering Science* 25: 1291–1300 <http://dx.doi.org/10.1089/ees.2007.0242> [accessed 17 July 2014].

Dustin, J.S. (2010) *Fundamentals of Operation of the Induced Bed Reactor (IBR) Anaerobic Digester*. PhD. Logan, Utah, US: Utah State University. <http://digitalcommons.usu.edu/cgi/viewcontent.cgi?article=1550&context=etd> [accessed 17 July 2014].

Eaton, A.D., Clesceri, L.S., Rice, E.W., Greenberg, A.E. and Franson, M.A.H. (2005) *Standard Methods for the Examination of Water and Wastewater*,

Centennial Edition 21st edn, Washington, D.C: American Public Health Association.

El-Mashad, H.E.-M.H., Zeeman, G., van Loon, W.K.P., Bot, G.P.A. and Lettinga, G. (2004) 'Effect of temperature and temperature fluctuation on thermophilic anaerobic digestion of cattle manure', *Bioresource Technology* 95: 191–201 <http://dx.doi.org/10.1016/j.biortech.2003.07.013> [accessed 17 July 2014].

El-Mashad, H.M. and Zhang, R. (2010) 'Biogas production from co-digestion of dairy manure and food waste', *Bioresource Technology* 101: 4021–4028 <http://dx.doi.org/10.1016/j.biortech.2010.01.027> [accessed 17 July 2014].

Finstein, M.S. (2010) 'Anaerobic digestion variants in the treatment of solid wastes', *Microbe* 5: 151–155 <http://dx.doi.org/10.1016/j.biortech.2003.07.013> [accessed 17 July 2014].

Fox, P. and Pohland, F.G. (1994) 'Anaerobic treatment applications and fundamentals: Substrate specificity during phase separation', *Water Environment Research* 66: 716–723 <http://www.jstor.org/stable/25044469> [accessed 22 July 2014].

Fry, J. (1975) *Methane Digesters for Fuel Gas and Fertilizers*, John Fry and Richard Merrill Santa Barbara, California <http://large.stanford.edu/courses/2010/ph240/cook2/docs/methane_digesters.pdf> [accessed 17 July 2014].

Fulford, D. (1988) *Running a Biogas Programme: A handbook*, London: Practical Action Publications (Intermediate Technology Publications). <http://development bookshop.com/running-a-biogas-programme-pb> [accessed 17 July 2014].

García-Gen, S., Lema, J.M. and Rodríguez, J. (2013) 'Generalised modelling approach for anaerobic co-digestion of fermentable substrates', *Bioresource Technology* 147: 525–533 <http://dx.doi.org/10.1016/j.biortech.2013.08.063> [accessed 17 July 2014].

García-Gen, S., Sousbie, P., Rangaraj, G., Lema, J.M., Rodríguez, J. and Torrijos, M. (2013) 'Disintegration and hydrolysis kinetics modelling for ADM1 application to codigestion: lab-scale model calibration with fruit and vegetable waste', *REDbiogàs*, Chile <http://www.redbiogas.cl/wordpress/wp-content/uploads/2013 July IWA-11928.pdf> [accessed 17 July 2014].

Griehl, C. and Hecht, C. (2009) 'Investigation of the accumulation of aromatic compounds during biogas production from kitchen waste', *Bioresource Technology* 100:654–658<http://dx.doi.org/10.1016/j.biortech.2008.07.034> [accessed 17 July 2014].

Harris, P. (2012) 'Anaerobic Digestion Model' [website -guide] Beginners Guide to Biogas <www.adelaide.edu.au/biogas/anaerobic_digestion/model/> [accessed 17 July 2014].

Hashimoto, A.G., Varel, V.H. and Chen, Y.R. (1981) 'Ultimate methane yield from beef cattle manure: Effect of temperature, ration constituents, antibiotics and manure age', *Agricultural Wastes* 3: 241–256 <http://dx.doi.org/10.1016/0141-4607(81)90011-1> [accessed 17 July 2014].

Hawkes, F.R., Guwy, A.J., Hawkes, D.L. and Rozzi, A.G. (1994) 'On-line monitoring of anaerobic digestion: Application of a device for continuous measurement of bicarbonate alkalinity', *Water Science and Technology* 30: 1–10 <www.iwaponline.com/wst/03012/wst030120001.htm> [accessed 22 July 2014].

Heeb, F. (2009) *Decentralised anaerobic digestion of market waste: Case study in Thiruvananthapuram, India*, Dübendorf, Switzerland: Swiss Federal Institute of Aquatic Science and Technology (Eawag). <www.eawag.ch/forschung/sandec/publikationen/swm/dl/Heeb_2009.pdf> [accessed 22 July 2014].

Hill, D.T. (1991) 'Steady-state mesophilic design equations for methane production from livestock wastes', *Transactions of the American Society of Agricultural Engineers* 34: 2157–2163 <http://agris.fao.org/agris-search/search.do?recordID=US9182207> [accessed 22 July 2014].

Hills, D.J. (1979) 'Effects of carbon: Nitrogen ratio on anaerobic digestion of dairy manure', *Agricultural Wastes* 1: 267–278 <http://dx.doi.org/10.1016/0141-4607(79)90011-8> [accessed 17 July 2014].

Husain, A. (1998) 'Mathematical Models of the Kinetics of Anaerobic Digestion: a Selected Review', *Biomass and Bioenergy* 14: p561–571 <9534(97)10047-2> [accessed 17 July 2014].

Huy, N.Q. (2008) *Sequential Dry Batch Anaerobic Digestion of the Organic Fraction of Municipal Solid Waste*. MSc. Bangkok, Thailand: Asian Institute of Technology, School of Environment, Resources and Development. <www.faculty.ait.ac.th/visu/Data/AIT-Thesis/Master Thesis final/Huy.pdf> [accessed 17 July 2014].

IEA (2001) *Biogas and More! Systems and Markets Overview of Anaerobic digestion*, IEA Bioenergy, Abingdon, UK. <www.biores.eu/docs/BIOGASFUNDAMENTALS/IEA-MSWAD.pdf> [accessed 17 July 2014].

Inglis, S.F., Gooch, C.A., Jones, L.R. and Aneshansley, D. (2007) *Cleanout of a plug-flow anaerobic digester after five years of continuous operation*, ASABE - Proceedings of the International Symposium on Air Quality and Waste Management for Agriculture, Broomfield, CO; United States. <800 53655815&partnerID=40&md5=f4743edf525c0af50e8a9574ee3eb68f> [accessed 17 July 2014].

Kaparaju, P., Buendia, I., Ellegaard, L. and Angelidakia, I. (2008) 'Effects of mixing on methane production during thermophilic anaerobic digestion of manure: Lab-scale and pilot-scale studies', *Bioresource Technology* 99: 4919–4928 <http://dx.doi.org/10.1016/j.biortech.2007.09.015> [accessed 17 July 2014].

Karim, K., Klasson, K.T., Drescher, S.R., Ridenour, W., Borole, A.P. and Al-Dahhan, M.H. (2007) 'Mesophilic digestion kinetics of manure slurry', *Applied Biochemistry and Biotechnology* 142: 231–242 <http://dx.doi.org/10.1007/s12010-007-0025-4> [accessed 17 July 2014].

Kusch, S., Oechsner, H. and Jungbluth, T. (2008) 'Biogas production with horse dung in solid- phase digestion systems', *Bioresource Technology* 99: 1280–1292 <http://dx.doi.org/10.1016/j.biortech.2007.02.008> [accessed 17 July 2014].

Lastella, G., Testa, C., Cornacchia, G., Notornicola, M., Voltasio, F. and Sharma, V.K. (2002) 'Anaerobic digestion of semi-solid organic waste: biogas production and its purification', *Energy Conversion and Management* 43: 63–75 <http://dx.doi.org/10.1016/S0196-8904(01)00011-5> [accessed 17 July 2014].

Lauwers, J., Appels, L., Thompson, I.P., Degrève, J., Van Impe, J.F. and Dewil, R. (2013) 'Mathematical modelling of anaerobic digestion of biomass and waste: Power and limitations', *Progress in Energy and Combustion Science* 39: 383–402 <http://dx.doi.org/10.1016/j.pecs.2013.03.003> [accessed 17 July 2014].

Lee, M., Hidaka, T. and Tsuno, H. (2008) 'Effect of temperature on performance and microbial diversity in hyperthermophilic digester system fed with kitchen garbage', *Bioresource Technology* 99: 6852–6860 <http://dx.doi.org/10.1016/j.biortech.2008.01.038> [accessed 17 July 2014].

Li, R., Li, X. and Chen, S. (2007) 'Performance evaluation of anaerobic digestion of dairy manure in plug flow reactor and continuous stirred tank reactor', *Nongye Gongcheng Xuebao/Transactions of the Chinese Society of Agricultural Engineering* 23: 186–190 <F134B> [accessed 22 July 2014].

Lidholm, O. and Ossiansson, E. (2008) *Modeling Anaerobic Digestion.* Master Thesis 2008-12. Lund, Sweden: Lund University, Sweden. <http://www.vateknik.lth.se/exjobbR/E539.pdf> [accessed 17 July 2014].

Liu, T. (1998) 'Anaerobic digestion of solid substrates in an innovative two-phase plug-flow reactor (tppfr) and a conventional single-phase continuously stirred-tank reactor' *Water Science and Technology* 38: p453–461 <http://dx.doi.org/10.1016/S0273-1223(98)00723-9> [accessed 17 July 2014].

Madegan, M.T., Martinko, J.M., Dunlap, P.V. and Clark, D.P. (2009) *Brock's biology of microorganisms* 12th edn, Benjamin Cummings, San Francisco, California. <www.pearsonhighered.com/educator/product/products_detail.page?isbn=0132324601> [accessed 17 July 2014].

Maramba, F.D. (1978) *Biogas and Waste Recycling: The Philippine experience,* Metro Manilla, Philippines: Maya Farms Division, Metro Flour Mills. <http://journeytoforever.org/biofuel_library/biogasPhilippines.pdf> [accessed 22 July 2014].

Martinez-Garcia, G., Johnson, A.C., Bachmann, R.T., Williams, C.J., Burgoyne, A. and Edyvean, R.G.J. (2007) 'Two-stage biological treatment of olive mill wastewater with whey as co-substrate', *International Biodeterioration & Biodegradation* 59: 273–282 <http://dx.doi.org/10.1016/j.ibiod.2007.03.008> [accessed 22 July 2014].

Mata-Alvarez, J., Macé, S. and Llabrés, P. (2000) 'Anaerobic digestion of organic solid wastes. An overview of research achievements and perspectives', *Bioresource Technology* 74: 3–16 <http://dx.doi.org/10.1016/S0960-8524-(00)00023-7> [accessed 17 July 2014].

Molinuevo-Salces, B., García-González, M.C., González-Fernández, C., Cuetos, M.J., Morán, A. and Gómez, X. (2010) 'Anaerobic co-digestion of livestock wastes with vegetable processing wastes: A statistical analysis', *Bioresource Technology* 101: 9497–9485 <http://dx.doi.org/10.1016/j.biortech.2010.07.093> [accessed 22 July 2014].

Nishio, N. and Nakashimada, Y. (2007) 'Recent development of anaerobic digestion processes for energy recovery from wastes', *Journal of Bioscience and Bioengineering* 103: 105–112 <http://dx.doi.org/10.1263/jbb.103.105> [accessed 17 July 2014].

Pavlostathis, S.G. (2011) *6.31 - Kinetics and Modeling of Anaerobic Treatment and Biotransformation Processes.* In: M. Moo-Young, ed. Comprehensive Biotechnology (Second Edition). pp. 385–397, Burlington: Academic Press. <http://dx.doi.org/10.1016/B978-0-08-088504-9.00385-8> [accessed 22 July 2014].

Reynell, C. (2008) 'Bioplex Technologies - Treating Waste Naturally' [website - presentation] <http://www.slideshare.net/nationalrural/swru022008-bioplex-technologies-feb08> [accessed 17 July 2014].

VFA in the system L.E., Boyle W.C. and Converse, J.C. (1986) 'Improved alkalimetric monitoring for anaerobic digestion of high-strength wastes', *Water Pollution Control Federation* 58: 406–411 <http://www.jstor.org/stable/25042933> [accessed 22 July 2014].

de Rosa, M. and Gambacorta, A. (1986) 'Structure, Biosynthesis, and Physicochemical Properties of Archaebacterial Lipids', *Microbiological Reviews* 50: 70–80 <www.ncbi.nlm.nih.gov/pmc/articles/PMC373054/> [accessed 22 July 2014].

Saravanan, V. and Sreekrishnan, T.R. (2006) 'Modelling anaerobic biofilm reactors - A review', *Journal of Environmental Management* 81: 1–18 <http://dx.doi.org/10.1016/j.jenvman.2005.10.002> [accessed 17 July 2014].

Seghezzo, L., Zeeman, G., van Lier, J.B., Hamelers, H.V.M. and Lettinga, G. (1998) 'A review: The anaerobic treatment of sewage in UASB and EGSB reactors', *Bioresource Technology* 65: 175–190 <http://dx.doi.org/10.1016/S0960-8524(98)00046-7> [accessed 17 July 2014].

Shah, O.P. (2006) *Sustainable Waste Processing in Mumbai: Using the Nisargruna Technology*. MSc Thesis. Borås, Sweden: Hogskolan i Borås. <http://www.kingdombio.co.uk/ShahThesis.pdf> [accessed 17 July 2014].

Smith, S.R., Lang, N.L., Cheung, K.H.M. and Spanoudaki, K. (2005) 'Factors controlling pathogen destruction during anaerobic digestion of biowastes', *Waste Management* 25: 417–425 <http://dx.doi.org/10.1016/j.wasman.2005.02.010> [accessed 17 July 2014].

Svoboda, I.F. (2003) *Anaerobic Digestion, Storage, Oligolysis, Lime, Heat and Aerobic Treatment of Livestock Manures*, Stoneleigh Park: FEC Services Ltd. <www.scotland.gov.uk/Resource/Doc/1057/0002224.pdf> [accessed 17 July 2014].

Uke, M.N. and Stentiford, E. (2013) 'Enhancement of the anaerobic hydrolysis and fermentation of municipal solid waste in leachbed reactors by varying flow direction during water addition and leachate recycle', *Waste Management* 33: 1425–1433 <http://dx.doi.org/10.1016/j.wasman.2013.02.020> [accessed 17 July 2014].

Vögeli, Y., Lohri, C.R., Gallardo, A., Diener, S. and Zurbrügg, C. (2014) *Anaerobic Digestion of Biowaste in Developing Countries: Practical Information and Case Studies*, Dübendorf, Switzerland: Swiss Federal Institute of Aquatic Science and Technology (Eawag). <www.eawag.ch/forschung/sandec/publikationen/swm/dl/biowaste.pdf> [accessed 17 July 2014].

Wang, Q., Kuninobu, M., Ogawa, H.I. and Kato, Y. (1999) 'Degradation of volatile fatty acids in highly efficient anaerobic digestion', *Biomass and Bioenergy* 16: 407–416 <http://dx.doi.org/10.1016/S0961-9534(99)00016-1> [accessed 17 July 2014].

Ward, A.J., Hobbs, P.J., Holliman, P.J. and Jones, D.L. (2008) 'Optimisation of the anaerobic digestion of agricultural resources', *Bioresource Technology* 99: 7928–7940 <http://dx.doi.org/10.1016/j.biortech.2008.02.044> [accessed 17 July 2014].

Wellinger, A. and Murphy, J. (2013) *Biogas Handbook: Science, production and application*, D Woodhead, Cambridge, UK Publishing Series in Energy. <http://www.iea-biogas.net/biogas-handbook.html> [accessed 17 July 2014].

Yen, H.-W. (2001) 'Anaerobic co-digestion of algal sludge and waste paper to produce methane', *Bioresource Technology* 98: 130–134 <http://dx.doi.org/10.1016/j.biortech.2005.11.010> [accessed 17 July 2014].

Zoetemeyer, R.J., Matthijsen, A.J.C.M., Cohen, A. and Boelhouwer, C. (1982) 'Product inhibition in the acid forming stage of the anaerobic digestion process', *Water Research* 16: 633–639 <http://dx.doi.org/10.1016/0043-1354(82)90084-7> [accessed 17 July 2014].

CHAPTER 5
Biogas effluent as compost

Abstract

The value of anaerobic digestion to produce compost has received little interest until recent years. Biogas effluent slurry, or bio-slurry, is increasingly seen as having value, not only as a fertilizer, but also in reducing crop disease and acting as a soil conditioner. The use of bio-slurry has lasting effects on the soil. The ways in which biogas effluent is processed affect the value of the product: composting and vermicomposting increase its usefulness. The benefits of bio-slurry are illustrated by the results from many practical examples.

Keywords: biogas, anaerobic digestion, slurry, compost

The history of using biogas slurry as compost

Anaerobic digestion is mainly used for three purposes, to:

- generate energy;
- process waste and remove obnoxious smells;
- generate compost.

The first two purposes have received most of the interest from academic researchers, but the value of the effluent as a compost has had much less attention. In recent years, the benefits of biogas effluent slurry, or *bio-slurry*, as it is sometimes called, have been recognized more widely.

In the Chinese programme, the benefits of using anaerobic digestion to produce compost were originally seen as more important than the gas produced. Chinese farming culture had traditionally recognized the need to recycle biomass materials to feed the soil in order to produce crops. In the traditional Chinese way of life, people were employed to collect human sewage each morning using 'honey wagons', so it could be placed on compost heaps at the edge of the village or town (White, 2005). In rural areas, most farms would breed pigs, not only for meat, but also as a way to recycle food residues, so that pig dung could be used to fertilize the crops.

The introduction of biogas plants into the ageold system of farming provided several strong benefits. Human sewage and pig dung could be fed directly into the plant, which removed the smell. Food wastes that were not fed to the pigs could be fed directly to the biogas plant, which behaved as a fast-acting compost heap (van Buren, 1979). However, this approach meant that the biogas plant had to be emptied every six months, so that the composted

http://dx.doi.org/10.3362/9781780448497.005

material could be removed and spread on the land. The fixed-dome biogas plants had a concrete plug in the top that could be removed to provide access to the digested material. The weakness of this approach was that every time this plug was removed and replaced, it became more difficult to seal it to retain the biogas. It also meant that the family could not use the gas for cooking until the plant had been refilled with slurry from the toilets and pigsties.

In the early Indian and Nepal programmes, the main emphasis was to use biogas to replace firewood for cooking, so the use of the effluent slurry received less publicity. Traditionally farmers collected their cattle dung and piled it up so that it could be spread on their fields before crops were planted. With a biogas plant, the dung is mixed with water before it is fed to the plant. The effluent slurry is collected in the same way as the dung had been, but it has to be left to dry in the sun, to remove the water, before it can be piled up to be spread on the fields.

The lack of emphasis on using the effluent slurry as a fertilizer meant that farmers have been receiving poor training on its use over the many years these programmes have been running. A visit to two neighbouring farmers in Nepal demonstrated the result. One farmer claimed that he had tried using the effluent slurry on a vegetable crop, cauliflowers, but it did not work well. The cauliflowers were not as good as when he had applied inorganic fertilizers. His neighbour said he had used the effluent slurry on a crop of cauliflowers and found they grew much better, growing to twice the size. The difference was in the way the farmers used the slurry. The young man on the second farm had received a good education and had explained to his father the best way to compost the biogas effluent with other biomass materials.

The biogas programmes in Asia supported by the Netherland Development Organisation (SNV) have placed greater emphasis on the use of the bio-slurry in recent years. In Bangladesh, the programme employs an agriculturalist to advise the extension workers, so they can teach the farmers about the benefits of the slurry (Islam, 2006). The biogas programme also contracted the government-run Bangladesh Agricultural Research Institute (BARI) to do research on the benefits of slurry (Noor and Sarker, 2012). In Vietnam, the biogas programme was encouraged by a local group, the Vietnam Gardening Association (VACVINA) (CCRD, 2010), to place emphasis on using biogas slurry as a fertilizer (von Eije, 2007). This work has been summarized by Hivos (Warnars and Oppenoorth, 2014).

Benefits of bio-slurry

There are three ways in which the slurry from an anaerobic digester can help the soil in which plants are grown. The immediate advantage is that it can provide essential nutrients to help crops grow. However, bio-slurry does much more than that.

Bio-slurry as a fertilizer

Anaerobic digestion does not add nutrients to the biomass that is placed in it. However, it does concentrate those nutrients. The main nutrients that are required by crops are nitrogen (N), phosphorous (P), and potassium (K).

There are also other elements that are required in smaller amounts, such as sulfur (S), calcium (Ca), magnesium (Mg). These are all present in most of the biomass materials that are added to a biogas plant, especially in animal dung. An anaerobic digester generates methane (CH_4) and carbon dioxide (CO_2), so removes carbon (C), hydrogen (H), and oxygen (O) from the slurry. This means that the proportion of useful crop nutrients increases in the slurry compared with the material that is fed in to the plant (de Groot and Bogdanski, 2013).

Anaerobic digestion also breaks down insoluble long-chain molecules into shorter molecules that dissolve in water and are more available to crops (Paulin and O'Malley, 2008). In most biomass, nitrogen is bound up in proteins and other plant constituents, which break down slowly in the soil. The microbes in a biogas plant convert proteins into a soluble form, such as ammonia, which is very easily absorbed by crop roots (Pandian, 2005). This is true of the other plant nutrients as well. However, some of the nutrients in the slurry that has immediately left the plant are not in a form that is useful. Sulfur, for example, is likely to be in the form of dissolved hydrogen sulfide, which is poisonous. If the slurry is left for a few days to mature, the hydrogen sulfide is oxidized and forms compounds that are more useful to crops.

Reduction of crop disease

The second main benefit of using slurry from an anaerobic digester is the microbes that grew in the digester. The methanogen microbes quickly become quiescent once the slurry is in the open air. However, many of the other microbes in the slurry are beneficial and assist in the breakdown of biomass materials in the soil and help release the nutrients that are bound in them. These microbes also compete with other, less helpful, microbes in the soil, such as crop pathogens (Satyamoorty, 1999). The addition of bio-slurry provides the existing soil microbes with reinforcements, increasing plant health and indirectly improving plants' defences.

Increased biological activity in the soil allows greater rates of breakdown of biomass materials in the soil. It also increases the rate of breakdown of any toxic chemicals in the soil, including those produced by pathogens. The increase in availability of plant nutrients also means that plants can grow more strongly and use their own defences against pests and diseases more effectively.

Enhancing soil structure

Practical experience of the use of bio-slurry is that the benefits increase as the bio-slurry is applied over a period of time, especially in the first three years. The reason for this can be understood by considering the recent enthusiasm for the use of biochar (Lehmann and Joseph, 2009). If biomass is heated in the absence of air – pyrolyzed in a biomass retort – volatile chemicals are released, leaving a carbon-rich char. If the heating is done slowly to a high temperature (600 °C and above), the process gives a fuel gas, mainly hydrogen and carbon monoxide, which can be used for heating or in engines, providing it is carefully cleaned. If the heating is done rapidly, to a slightly lower temperature, the process produces a range of flammable liquids that can be used as a fuel.

The char that remains from pyrolysis has been shown to enhance fertility in soils (Bruges, 2009). It has been suggested that this biochar (Terra Preta) (Wayne, 2012) was used many years ago in parts of Latin America to build up the soil to grow crops for ancient civilisations. Biochar is inert and does not supply any crop nutrients. Lignin, in biomass, acts as a scaffold at the microscopic level, that supports other biomass constituents, such as cellulose and hemicellulose (Abdul Khalil *et al.*, 2006). The lignin molecule is much stronger than the other constituents, so its structure holds together when the biomass is pyrolyzed. The remaining carbon is therefore in the form of a microscopic matrix, which enables it to act as a sponge, absorbing and retaining moisture and plant nutrients in the top layer of the soil. Biochar prevents leaching, the transfer of plant nutrients to lower layers, as water percolates through the soil. Moisture and crop nutrients are therefore kept in reach of the roots of crops.

Biological degradation of biomass works in a similar way. Both anaerobic and aerobic digestion will preferentially break down the less resilient molecules, such as cellulose and hemicellulose, so the lignin retains its structure. Bio-slurry contains this lignin, which forms an essential part of the humus in the soil and acts as a sponge to retain moisture and crop nutrients in the top layer of the soil. The regular addition of bio-slurry to the soil allows the amount of humus to increase, thus enhancing the soil's ability to retain moisture and dissolved crop nutrients.

There are differences between biochar and bio-slurry compost that are listed in Table 5.1. Biochar cannot be degraded in the soil, so remains as a permanent part of the soil structure, whereas the lignin in bio-slurry will slowly degrade over time. Biochar does not introduce extra microbes into the soil, although it will act as a support medium for those that are already there. Bio-slurry adds extra beneficial microbes to reinforce the ones that are already there. These extra microbes are responsible for the slow break down of the lignin in the humus layer. Biochar has within it active sites that allow it to adsorb organic chemicals.

Table 5.1 Differences between biochar and bio-slurry

Feature	Biochar	Bio-slurry
Degradation	Remains in soil for thousands of years	Degrades over time (several years)
Microbes	Supports existing microbes	Adds beneficial microbes to soil
Nutrients	Retains existing and added nutrients	Adds and retains nutrients
Leaching	Resists leaching of nutrients	Resists leaching of nutrients
Moisture	Retains moisture in soil	Retains moisture in soil
Toxins	Binds toxins	Toxins slowly broken down

It is similar to the action of the activated charcoal that is used in filters to remove obnoxious smells and toxic gases from air and water.

The microscopic sponges in the humus from bio-slurry also contain active sites that bind organic chemicals, including toxins. Microbial activity breaks down organic chemicals, including toxins, and releases nutrients into the soil.

Biochar must be used with the addition of inorganic fertilizers because it does not contain any food nutrients. Bio-slurry compost does contain plant nutrients, but the addition of inorganic fertilizers will enhance its effect on crop production, especially when it is first used. The ability of both to retain plant nutrients and prevent leaching means that much less inorganic fertilizer is required. As bio-slurry compost is added over several years, the amount of inorganic fertilizer required is reduced over time.

Processing of bio-slurry

The way that the bio-slurry is treated after it comes from a biogas plant as effluent affects its value as a fertilizer. Biogas plant effluent contains the components that make a good bio-slurry and these can be improved with processing.

Traditional approaches

In the early biogas programmes in India and Nepal, farmers used the biogas plant effluent in the same way that they had used the animal dung that they put into the plant. The slurry was collected in a pond and allowed to dry in the sun. As the dried slurry accumulated, it was raked to the side of the pond and dug out to add to a heap of compost. When the compost was needed for crops, it could be dug out of the pile and carried to the fields in baskets.

Allowing the slurry to dry in the sun ensured that it was mature, that the poisonous hydrogen sulfide had evaporated or been oxidized and any remaining volatile fatty acids had also decomposed. However, sun drying also meant that much of the available nitrogen in the bio-slurry was lost (Pandian, 2005). The breakdown of proteins and other nitrogen containing food materials forms ammonia, which dissolved in the water to form ammonium hydroxide. As the water is evaporated by the heat of the sun, the ammonia also evaporates (Le, 2008). As a gas in the air, it is quickly oxidized to water and nitrogen and lost as a crop nutrient.

In Bangladesh, farmers are encouraged to build simple roofs over their slurry pits, made from wood that is thatched with straw. This prevents the slurry from drying out too fast and allows a greater amount of the ammonia to remain in the slurry. In Vietnam, farmers are encouraged to spread the liquid slurry over the fields using watering cans (Ngo, 2010). This works well, as long as the slurry is left for a few days for the hydrogen sulfide to evaporate or oxidize.

Bio-slurry and compost

An improvement on the traditional approach, encouraged by SKG Sangha, is to place dry biomass in the pond into which the effluent slurry flows (Ashden, 2007). This material, which can be straw, dry grasses, cut weeds, or dried leaves, will absorb the liquid in the slurry, preventing it from evaporating. The dissolved nitrogen is also absorbed and prevented from evaporating. In many rural households, grass is used to make mats and other household items, which have a limited lifetime, so such dry material is usually easily available. The microbes in the slurry accelerate the composting process of the dry biomass.

The wetted material should be regularly dug from the slurry pond and placed in a compost heap, so the dry material in the slurry pit can be replaced (VK NARDEP, 1996). The microbes from the slurry will begin to break down the dry biomass to form good quality compost. This composting process is usually twice as fast as that in a conventional compost heap, because the right types of microbes are already available in the slurry and ready to grow. The presence of lignin in the dry biomass means that the compost contains a greater proportion of humus material than normally exists in the effluent bio-slurry. It also means that the compost is fairly dry and can be dug up and carried to the fields in a basket.

The compost needs to be piled up to form heaps that are at least 2 m in diameter and 1 m high, so that the centre can increase in temperature and form the compost. Larger heaps can be formed into a windrow, a long heap that is about 2 m wide, providing there is space and compost available. In order to make good compost, the heap should be turned over at least once a week, to ensure the aerobic microbes have sufficient air. The centre of a heap will heat up, as energy is released by the breakdown processes, and then cool down again. The compost is ready for use once the whole heap has cooled down.

Vermi-composting

One step that further improves the compost is to use vermi-composting. Once a compost heap has matured and cooled down to ambient temperature, it can be fed to earthworms (Gajalakshmi and Abbasi, 2004). A vermi-composting unit can be a container into which the compost and worms are placed. The worms live in the top layer, but need to be sheltered from the sun and kept moist by being regularly sprinkled with water. The top layer of compost can be carefully removed each morning, as the worms burrow deeper into the soil exposed to light. Once the amount of compost in the bin has been reduced, the worms can be moved to another bin and the first bin refilled with compost from the matured compost heap.

Vermi-compost has all of the benefits of bio-slurry, but they are further enhanced (Edwards *et al.*, 2010). SKG Sangha in South India (Ashden, 2007) is building systems that allow farmers to use vermi-compost made from compost generated with effluent slurry from a biogas plant (Figure 5.1). Farmers who have used this material over three years claim that production of vegetables and other crops has an increased production by one third compared with when they did not use vermi-composting. They also claim that they can sell their crops in the local markets at a price one third more than those grown by other local farmers because of the improvement in quality.

Figure 5.1 Couple with biogas plant in Nepal

If farmers wish to sell some of their vermi-compost, the demand appears to be very high. The owners of palm tree plantations, such as coconut and areca nut, are keen to use vermi-compost, because they claim it reduces the number of immature nuts that fall from the trees (Ashden, 2007). Other crops that are grown with vermi-compost have a reduced incidence of pests and diseases (Arancon and Edwards, 2005). If excess moisture is drained from the base of the vermi-compost pit, using a tap, the 'compost tea' is a valuable foliar feed, which also acts to control pests and diseases (Edwards and Arancon, 2010).

The balance of materials in the compost that is fed to the worms needs to be correctly managed. An excess of dry biomass, containing fibrous lignin, can prevent the worms breaking down the mixture. This ligninaceous material can be broken down by the addition of fungal spores to the compost heap before or after the main composting process. If these spores are from edible fungi, such as mushrooms, an extra valuable crop can also be harvested.

Bio-slurry and sanitation

There is a concern about using compost on crops that it might contain human or animal pathogens. If the feed material used in the digester includes animal dung or human sewage, pathogens could pass through and be present in the compost. There are also cultural concerns; people from some cultures do not like the 'idea' of using compost from human sewage, even when it has been properly digested.

Cultural attitudes to bio-slurry and sanitation

In China, the use of night soil as a fertilizer has long traditions, so the incorporation of a latrine into a biogas system is acceptable. People have found that basic composting does reduce the potential for the transfer of disease from human waste. Research has shown the value of biogas digestion in inactivating and removing viable ova of gut parasites from night soil (Remais *et al.*, 2009).

In Nepal and India, in areas in which Hinduism is the main religion, cattle dung is seen in a positive light, because cows are holy animals. A biogas plant fed with cattle dung is therefore strongly culturally appropriate, so people are keen to use both the gas and the effluent slurry. Attaching a latrine to a biogas plant can therefore have negative cultural effects, as people may be less willing to use the gas for cooking and especially be resistant to using the compost. However, such attitudes can change with time. A survey of biogas plants in Nepal, done in 1991 (Fulford *et al.*, 1991), when latrines were being added to biogas plants under the United Nations Development Programme (UNDP) funded programme, showed that the older generation were unwilling to use the latrines. However, when their children were questioned on their own, they said they preferred using the latrine in the yard to using the fields. During a field visit in 2012 (Fulford *et al.*, 2012) an elderly couple explained how enthusiastic they were about their latrine, which improved sanitation on their farm (Figure 5.2). In Nepal, attitudes had changed over 20 years partly through extensive publicity.

Attitudes remain varied in Ghana (Bensah and Brew-Hammond, 2008), with some groups being unwilling to use bio-slurry from the processing

Figure 5.2 Biogas sewage plant in a jail in Cyangugu in Rwanda

of sewage, while other groups were keen on the benefits of improved crop production. The bio-slurry from the processing of sewage from jails in Rwanda (Ashden, 2005) is used to fertilize crops and also to enable flowers to be grown in attractive gardens (Figure 5.3).

Figure 5.3 Vermi-compost plant in South India

Actual effects of bio-slurry and sanitation

Tests on pathogens in bio-slurry taken directly from anaerobic digesters demonstrated that some pathogen species (such as Salmonella) are reduced by the process of digestion, while others remain viable (such as Listeria) (de Groot and Bogdanski, 2013). Tests on bio-slurry that had been allowed to mature (Alfa *et al.*, 2014) show that some pathogens (e.g. Salmonella) do exist in the compost, although the numbers had decreased by well over 99.9 per cent. Pathogens that prefer an aerobic environment will be under pressure in the digester, while those that prefer an anaerobic environment will thrive. However, if the slurry is then composted in an aerobic environment, the anaerobic microbes will come under pressure. Also, a compost heap will increase in temperature to above 60 °C, so many more pathogens will be killed. The time spent in composting (several weeks) will remove most viruses because they need host cells to survive. The use of vermi-composting will further destroy pathogenic microbes because worms digest microbes in their guts (Nagavallemma *et al.*, 2004).

The application of vermi-compost on crops does reduce the incidence of plant diseases and pests (Edwards *et al.*, 2010). Grapes and strawberries that were grown on plots with vermi-compost added were significantly less affected by common plant diseases than those on control plots. Tests were done in which vermi-compost was sterilized by heating and applied to plots alongside other plots of the same crop to which non-sterilized vermi-compost had been added. The plots with sterilized vermi-compost showed more damage by disease, which suggested that reduction of disease was caused by competition from beneficial microbes in the vermi-compost.

Benefits are also gained by making a compost tea from soaking vermi-compost in water. The use of this compost tea as a foliar feed was shown to reduce the damage done by pests such as aphids and caterpillars, as well as reducing the effects of parasitic nematodes and fungi (Edwards and Arancon, 2010). Worms appear to produce enzymes that are toxic to such pests and these enzymes can be washed from their excretions.

Practical experience with bio-slurry, compost and vermi-compost

There is a growing number of studies that have been done with bio-slurry, compost and vermi-compost. There do not seem to be any academic studies on the process that uses bio-slurry, mixed with dry biomass to make compost, which is then further processed by vermi-composting. However, users of the results of this whole process, using systems built by SKG Sangha in South India (Ashden, 2007), do claim that their crop productivity (especially for vegetables, such as carrots, and spices, for example ginger) increases by at least one third and that they can sell their crops at a price that is one third more in the local markets.

Measurements of the percentage of nitrogen and potassium in dung and slurry in Bangladesh (Noor and Sarker, 2012) suggested that the nitrogen value was slightly reduced in cattle dung slurry (a mean of 1.14 per cent compared to 1.21 per cent in the manure), while the potassium was increased (a mean of 0.31 per cent over 0.25 per cent for the manure). In Bangladesh, the slurry is dried before use, so this result confirms that nitrogen is lost by drying the slurry. In poultry manure, both the nitrogen and potassium increase (1.56 per cent for nitrogen compared to 1.51 per cent in manure and 0.4 per cent for potassium as compared to 0.31 per cent in the manure). However, the use of bio-slurry on crops gives a crop increase of up to 254 per cent over an unfertilized plot for broccoli and even 26 per cent over a plot using a standard amount of inorganic fertilizer (Noor and Sarker, 2012). The figures for tomatoes were a 276 per cent increase over an unfertilized plot and 23 per cent over a plot using a standard dressing of inorganic fertilizer.

Tests in Vietnam with the use of liquid bio-slurry on crops suggest crop production improvements of 22 per cent for soya bean and 25.8 per cent for maize against crop from plots on which no fertilizer was used. When bio-slurry is used with inorganic fertilizer for cabbage, the increase was 24 per cent compared with a plot on which only inorganic fertilizer was used (Le, 2008). However, the increases for buckwheat and peanuts were only about 8 per cent. In Vietnam, the main feed material for a biogas plant is pig dung, which is much more liquid than cattle dung. The bio-slurry is also liquid, although it has a much reduced smell than the original dung. The bio-slurry is kept in ponds to mature and not allowed to dry. The liquid bio-slurry is poured over the crops from watering cans.

Another set of trials in Vietnam (von Eije, 2007) gave an increase of 29 per cent for a crop of tomatoes when bio-slurry was used with inorganic fertilizer, even when a reduced amount of inorganic fertilizer was applied. The amount of pesticides was also reduced, as the crop was less affected by pests. Commercial tea growers in a hill province of Vietnam who owned biogas plants were questioned. They were using liquid bio-slurry combined with inorganic fertilizer. They claimed a much improved income because the amount of chemical fertilizer they required was reduced, in addition they did they have to use as much pesticide to reduce the damage caused to the leaves. The tea crop grew faster, so more could be harvested. The quality of the tea was also increased, so tea traders were prepared to pay a high price for the crop. The value of the benefits from the use of bio-slurry was so high that the farmers claimed that they recovered the cost of the biogas plant within one year, even if the value of the biogas was ignored.

Trials were made with a potato crop in the Andes in Peru, comparing the use of liquid bio-slurry and composted manure against a control plot to which neither had been added (de Groot and Bogdanski, 2013). The bio-slurry gave a 27.5 per cent improvement in yield, compared to 15.1 per cent when composted manure was applied. In India, trials were done to compare the yields from sugar cane fertilized with vermi-compost, bio-slurry, animal manure and inorganic fertilizer against those from unfertilized plots (Singh *et al.*, 2007). The increases were 45 per cent for vermi-compost, 36 per cent with bio-slurry, and 34 per cent for animal manure, as against 44 per cent when an optimum amount of inorganic fertilizer was used.

The value of compost in crop production has been recognized for centuries, although a recent emphasis on the crop nutrient content has meant that its value as a soil conditioner has been neglected. The increased crop production from vegetables grown on plots of land which have been treated with compost is being recognized, for example, a 31 per cent increase in the yield of broccoli, especially when the amount of inorganic nitrogen was reduced (Paulin and O'Malley, 2008). The quality of the crop is also seen to increase. For example, the yield of marketable carrots was 43 per cent higher when the soil was treated with more than two times the quantity of compost, while the inorganic nitrogen was reduced by over 200 per cent. The use of compost as a mulch has a further benefit, it conserves moisture in the soil.

Vermi-composting is seen to provide a better result than composting biomass materials in a pile or windrow (Edwards and Arancon, 2010). The C:N ratio is reduced and vermi-compost is seen to contain plant-growth-enhancing hormones (Nagavallemma *et al.*, 2004). The use of vermi-compost on tomato crops in central India increased production by 66 per cent over controls as well as improving the quality of the crop. Vermi-compost retains nitrogen and moisture in the soil and reduces the amount of nitrogen that leaches out of the soil (Gajalakshmi and Abbasi, 2004). The production from tomatoes grown in a greenhouse in pots filled with a mixture of 30 per cent vermi-compost and a peat-based growth medium increased by 61 per cent over those grown in the medium alone (Arancon and Edwards, 2005). However, increased percentages of vermi-compost in the mix produced lower yields of tomatoes. For green peppers grown in pots filled with the same mixture, the optimum mix contained 40 per cent vermi-compost and gave a 72 per cent increase in marketable yield over those grown in the medium by itself. Again, increased proportions of vermi-compost decreased the yield. Field trials on grapes showed an increase in productivity of 55 per cent, with no loss of flavour.

The other benefits of vermi-composting that have been measured are the effects on populations of pests as well as beneficial organisms in the soil. Different amounts of vermi-compost were applied to plots on which grapes were grown in the USA (Arancon *et al.*, 2003). Samples of soil were extracted from around the roots of plants in each plot and the numbers of parasitic nematodes were counted in each 20 g soil sample. The numbers of parasitic nematodes was reduced by up to 71 per cent compared to samples taken from the roots of plants grown on plots fertilized with inorganic fertilizers. Similar trials with strawberries showed a reduction in parasitic nematodes of up to 58 per cent. Populations of beneficial organisms increased; the number of nematodes that consume fungi rose by up to 62 per cent for the grape plots and 382 per cent for the strawberry plots.

Tests on different approaches to composting have demonstrated that aerobic composting followed by vermi-composting give higher reductions in the amount of solids in the original biomass material (sewage sludge mixed with paper wastes) than either aerobic composting or vermi-composting on their own (Ndegwa and Thompson, 2001). A system that uses aerobic composting followed by vermi-composting gives a 45 per cent reduction in total solids, compared to 35.5per cent using vermi-composting alone. The volatile solid content was reduced by 13.4 per cent using the two-stage process, compared to 10.3 per cent by vermi-composting alone. The two-stage process also allowed the temperature of the composting system to increase to over 55 °C for three days, which means that most pathogenic microbes were likely to be killed. A vermi-composting system cannot be used to pasteurize the biomass because the temperatures would be too high for the worms to survive. Other tests on a similar two-stage process (Nair *et al.*, 2006) confirm that several common pathogens were strongly reduced, but the product needed to be vermi-composted for over two and a half months to ensure it was within international standards.

The recommended approach of anaerobic digestion followed by aerobic composting and then vermi-composting offers the benefits of all these processes. Further tests need to be done to obtain definite figures.

References

Alfa, M.I., Adie, D.B., Igboro, S.B., Oranusi, U.S., Dahunsi, S.O. and Akali, D.M. (2014) 'Assessment of biofertilizer quality and health implications of anaerobic digestion effluent of cow dung and chicken droppings', *Renewable Energy* 63: 681–686 <http://dx.doi.org/10.1016/j.renene.2013.09.049> [accessed 21 July 2014].

Arancon, N.Q. and Edwards, C.A. (2005) 'Effects of Vermicomposts on Plant Growth', in *International Symposium Workshop on Vermi Technologies for Developing Countries*, Los Baños, Philippines. <http://www.slocountyworms.com/wp-content/uploads/2010/12/EFFECTS-OF-VERMICOMPOSTS-ON-PLANT-GROWTH.pdf> [accessed 21 July 2014].

Arancon, N.Q., Galvis, P., Edwards, C. and Yardim, E. (2003) 'The trophic diversity of nematode communities in soils treated with vermicompost: The 7th international symposium on earthworm ecology · Cardiff · Wales · 2002', *Pedobiologia* 47: 736–740 <http://dx.doi.org/10.1078/0031-4056-00752> [accessed 21 July 2014].

Ashden (2005) 'Kigali Institute of Science, Technology and Management (KIST), Rwanda', [website - case study] Ashden <www.ashden.org/winners/kist05> [accessed 21 July 2014].

Ashden (2007) 'SKG Sangha, India, Biogas for cooking plus fertiliser from slurry', [website - case study] Ashden <www.ashden.org/winners/skgsangha> [accessed 21 July 2014].

Bensah, E.C. and Brew-Hammond, A. (2008) *Biogas Effluent and Food Production in Ghana*, Kumasi, Ghana: Kwame Nkrumah University of Science and Technology. <http://energycenter.knust.edu.gh/downloads/6/64.pdf> [accessed 21 July 2014].

van Buren, A. (1979) *A Chinese Biogas Manual*, London: Practical Action Publications (Intermediate Techonology Publicatons). <http://developmentbookshop.com/chinese-biogas-manual-pb> [accessed 22 July 2014].

CCRD (2010) *Biogas Training Material: Improved VACVINA model*, Hanoi, Vietnam: The Center for Rural Communities Research and Development (CCRD). <content/uploads/2010/10/Biogas-training-material-for-technicians-and-users.pdf> [accessed 21 July 2014].

Edwards, C.A. and Arancon, N.Q. (2010) *The Science of Vermiculture: The use of earthworms in organic waste managements*, Columbus, Ohio, USA: Soil Ecology Laboratory, Ohio State Univesrity. <www.slocountyworms.com/wp- content/uploads/2010/12/THE-SCIENCE-OF-VERMICULTURE.pdf> [accessed 21 July 2014].

Edwards, C.A., Arancon, N.Q. and Sherman, R.L. (2010) *Vermiculture Technology: Earthworms, organic wastes, and environmental management*, CRC Press, Boca Raton, FL USA.

von Eije, S. (2007) *Dong for Dung: The economic impact of using bioslurry for tea production on a household level in Thai Nguyen Province*, Vietnam, Hanoi: SNV and BPD. <www.snvworld.org/files/publications/economic_impact_of_bioslurry_for_production_in_thai_nguyen_province_vietnam_2007.pdf> [accessed 21 July 2014].

Fulford, D., Devkota, G.P. and Afful, K. (2012) *Evaluation of Capacity Building in Nepal and Asia Biogas Programme*, Hanoi, Vietnam: Kingdom Bioenergy Ltd for SNV. <www.kingdombio.com/Final Report - whole.pdf> [accessed 21 July 2014].

Fulford, D., Poudal, T.R. and Roque, J. (1991) *Evaluation of On-going Project: Financing and construction of biogas plants*, United Nations Capital Development Fund, New York.

Gajalakshmi, S. and Abbasi, S.A. (2004) 'Earthworms and vermicomposting', *Indian Journal of Biotechnology* 3: 486–494 <http://nopr.niscair.res.in/handle/123456789/5894> [accessed 22 July 2014].

de Groot, L. and Bogdanski, A. (2013) *Bioslurry = Brown Gold? A review of scientific literature on the co-product of biogas production*, Rome: FAO. <www.fao.org/docrep/018/i3441e/i3441e.pdf> [accessed 21 July 2014].

Islam, M.S. (2006) 'Use of Bioslurry as organic fertilizer in Bangladesh agriculture', in *International Workshop on the Use of Bioslurry Domestic Biogas Programmes*, Bangkok, Thailand: Grameen Shakti. <www.snvworld.org/sites/www.snvworld.org/files/publications/bangladesh_bio-slurry_use_in_agriculture_2006.pdf> [accessed 21 July 2014].

Khalil, H.P.S.A., Siti, M.A. and Mohd Omar, A.K. (2006) 'Chemical composition, anatomy, lignin distribution, and cell wall structure of Malaysian plant waste fibers', *BioResources* 1: 220–232 <http://ojs.cnr.ncsu.edu/index.php/BioRes/article/view/BioRes_01_2_220_232_AbdulKahlil_SM_Chemical_Composition_Mayalsian_Plant_Fibers> [accessed 22 July 2014].

Le, T.X.T. (2008) *Bio-slurry utilization in Vietnam*, Hanoi, Vietnam: The Biogas Program for the Animal Husbandry Sector of Vietnam, MARD Vietnam. <http://biogas.org.vn/vietnam/getattachment/Thu-vien-phim,-anh/An-pham/Bao-cao-tinh-hinh-su-dung-phu-pham-khi-sinh-hoc-ta/Bao-cao-tinh-hinh-su-dung-phu-pham-khi-sinh-hoc-tai-Viet-Nam.pdf.aspx> [accessed 21 July 2014].

Lehmann, J. and Joseph, S. (2009) *Biochar for Environmental Management: Science and technology*, Earthscan Abingdon, UK.

Nagavallemma, K.P., Wani, S.P., Lacroix, S., Padmaja, V.V., Vineela, C., Babu Rao, M. and Sahrawat, K.L. (2004) *Vermicomposting: Recycling wastes into valuable organic fertilizer*, Global Theme on Agrecosystems Report no. 8. Patancheru 502 324, Andhra Pradesh, India: International Crops Research Institute for the Semi-Arid Tropics. <http://ejournal.icrisat.org/agroecosystem/v2i1/v2i1vermi.pdf> [accessed 21 July 2014].

Nair, J., Sekiozoic, V. and Anda, M. (2006) 'Effect of pre-composting on vermicomposting of kitchen waste', *Bioresource Technology* 97: 2091–2095 <http://dx.doi.org/10.1016/j.biortech.2005.09.020> [accessed 21 July 2014].

Ndegwa, P. and Thompson, S. (2001) 'Integrating composting and vermicomposting in the treatment and bioconversion of biosolids', *Bioresource Technology* 76: 107–112 <http://dx.doi.org/10.1016/S0960-8524(00)00104-8> [accessed 21 July 2014].

Ngo, Q.V. (2010) *Utilization of Liquid Bio-Slurry as Fertilizer for Green Mustards and Lettuces in Dong Nai Province*, Vietnam: Institute of Agricultural Science for Southern Vietnam. <www.snvworld.org/files/publications/utilisation_of_bio-slurry_as_fertiliser_for_green_mustards_and_lettuces.pdf> [accessed 21 July 2014].

Noor, S. and Sarker, M.J.U. (2012) 'Bioslurry Presentation' [Presentation] Dacca, Bangladesh Agricultural Research Institute (BARI).

Pandian, S.P. (2005) *Biogas Manure User's Guide*, Kanyakumari, South India: Vivekananda Kendra – Nardep.

Paulin, B. and O'Malley, P. (2008) *Compost Production and Use in Horticulture*, Western Australia: Department of Agriculture and Food. <www.agric.wa.gov.au/objtwr/imported_assets/content/hort/compost_bulletin08.pdf> [accessed 21 July 2014].

Remais, J., Chen, L. and Seto, E. (2009) 'Leveraging rural energy investment for parasitic disease control: Schistosome ova inactivation and energy co-benefits of anaerobic digesters in rural China', *PLoS ONE* 4 <http://dx.doi.org/10.1371/journal.pone.0004856> [accessed 21 July 2014].

Satyamoorty, K. (1999) *Biogas - a Boon, Handbook on Biogas*, Kanyakumari, South India: Vivekananda Kendra – NARDEP. <http://www.vknardep.org/publications/english-books/189-biogas-a-boon.pdf> [accessed 21 July 2014].

Singh, K.P., Suman, A., Singh, P.N. and Srivastava, T.K. (2007) 'Improving quality of sugarcane-growing soils by organic amendments under subtropical climatic conditions of India', *Biology and Fertility of Soils* 44: 367–376 <http://dx.doi.org/10.1007/s00374-007-0216-8> [accessed 21 July 2014].

VK NARDEP (1996) *Biogas Slurry and Farmyard Manure*, Kanyakumari, South India: Vivekananda Kendra – NARDEP.

Warnars, L. and Oppenoorth, H. (2014) *Bioslurry, a Supreme Fertilizer: A study on bioslurry results and uses*, Hivos, The Hague, Netherlands<http://hivos.org/sites/default/files/bioslurry_book.pdf> [accessed 21 July 2014].

Wayne, E. (2012) *Conquistadors, Cannibals and Climate Change: A brief history of biochar*, Paris: Pro-Natura International. <www.pronatura.org/wp-content/uploads/2013/02/History-of-biochar.pdf?PHPSESSID=4c5981ce27a45ffdec1cf4c1188c15e7> [accessed 21 July 2014].

White, R. (2005) *The Role of Biogas in Rural Development and Resource Protection in China: A case study of Lijiang Municipality*, Yunnan Province, China: National Science Foundation and Michigan State University <htttp://amazonaws.com/zanran_storage/forestry.msu.edu/ContentPages/16804398.pdf> [accessed 22 July 2014].

CHAPTER 6
Main domestic biogas plant designs

Abstract

From the wide range of biogas digester designs, only a few meet the criteria for use in the programmes in Asia which have been made in large numbers. The floating drum system was developed in India, but has largely been replaced by fixed-dome systems. Fixed-dome systems were developed in China and the two main approaches are to use a dome made from brick or concrete. There are several aspects that are common to both fixed-dome designs: such as how the dome is sealed, how reservoir pits are made, the covers over reservoir pits and inlet pits. Various attempts have been made to adapt such plants for climates that have lower ambient temperatures.

Keywords: biogas, anaerobic digestion, plant designs

Background

There is a very wide range of digester designs in the literature (Rajendran *et al.*, 2012). Since biogas technology has been used widely for over 30 years, several basic designs have stood the test of time, especially those built in the programmes of China, India, and Nepal. Even these programmes started badly as the innovators in the different countries attempted to find a design that was both cheap and easy to make, while also being reliable and long lasting. People were also able to learn from each other, as they were prepared to share their experiences and publish these designs. Persistence, the sharing of ideas, and the use of trial-and-error eventually led to several basic designs that have enabled millions of units to be built at prices that are affordable and have a long lifetime. While a few of the early plants made to these designs did fail, as people learned how to design and make these plants effectively, there are still very large numbers of plants made to these designs which are still working 20 to 30 years after they were built.

The potential lifetime of a brick or concrete dome can be illustrated by similar structures that have been made in the past. The Gol Ghar *godown* (storage unit) in Patna, India (Sharma, 2013) was built in 1786 by Captain John Garstin. It has a diameter at the base of 125 m and is 29 m high. It is made in approximately the same shape as a brick-dome digester and is still strong enough to allow tourists to climb to its top. One of the best examples of a concrete dome structure that has survived for a very long time is the rotunda of the Pantheon in Rome (Mark and Hutchinson, 1986), which was built in about 128 AD and has lasted over 1880 years. This dome spans a diameter of 43.4 m and is 21.7 m high, above a 21.7 m cylindrical base.

http://dx.doi.org/10.3362/9781780448497.006

Most of the plant designs that have been used in the programmes of China, India, and Nepal are appropriate for sizes that can be mainly used for domestic purposes, usually supplying cooking fuel for individual families. The original digester pit volumes ranged from 7 m^3 to 20 m^3. Increasing experience with these plants (Tuladhar *et al.*, 1985; VK NARDEP, 1993) suggested that people could get sufficient daily gas from smaller, cheaper digester volumes down to 3 m^3, if animal dung is used as the feed, and 1 m^3, if food and agricultural residues are used (Ashden, 2007).

The more recent interest in using biogas technology to process larger quantities of residues, for example those from food processing, has led to a requirement for larger digester volumes. In Europe and the USA, the demand has been for systems in a central place that can process food and other wastes collected from domestic and commercial sources. Over the past 120 years, sewage sludge has been collected and processed in anaerobic digesters with volumes of 1,000 m^3 and above. These designs are relatively expensive, because they require the use of tanks made of steel or concrete with steel reinforcement, but the low-cost small-scale approach, as used in Asia, has never seemed relevant in this context.

However, the few people who have set up successful projects in Africa have been those that have adapted the Asian ideas and applied them to their own conditions. These innovators have been able to make individual digester tanks of up 100 m^3 in volume using Asian technology to process sewage sludge from institutions such as hospitals, prisons, schools, universities, and hotels (Ashden, 2005).

Digester design criteria

The demands of the microbes that are involved in anaerobic digestion determine a set of basic functions that a digester has to perform. The system must contain a volume of slurry for a suitable length of time (usually several days); it must keep the slurry at a reasonable temperature, with the minimum of fluctuations; it must keep out air and light; and it must collect and store the biogas generated for a period of time. The slurry, while it is being digested, contains volatile fatty acids which are strongly polluting to groundwater, so there must be no leakages of liquid into the surrounding soil.

A biogas plant needs to fulfil of the following criteria:

- able to contain the slurry without leaking;
- leak-tight against both slurry and gas;
- cheap to build, using local materials as far as possible;
- easy to use and maintain;
- easy to insulate and heat (if ambient temperatures are low);
- reliable and free from possible failures.

For tanks to contain a large volume of slurry, they need to be sufficiently strong. Even a small digester of only 1 m^3 of volume and 1 m high needs to contain a pressure of 10 kPa at the base without leaking. Most tanks are cylindrical or partially spherical, as rectangular tanks have flat sides which are likely to flex and corners which tend to crack and leak when liquid is placed in them. The outside walls of the tank are under tension because the pressure is outwards. Tanks built above ground must be properly reinforced in order to withstand the pressure. Tanks are usually made of steel or have steel reinforcement in the structure. Small tanks can be made of plastic, but need to be manufactured with reinforcement flanges to resist flexing (see Chapter 7).

Low-cost plants are usually made from masonry – either concrete or brick. They are built underground so that the surrounding earth supports the tank walls. If bricks are used, they need to be made from fired clay and be of reasonable quality. Bricks are usually covered with cement plaster because it reduces the porosity of the brick and stops slurry or gas leaking out.

Concrete is a good material for making biogas plants – it can be cast into the right shapes using formwork. As long as the concrete is always under compression, it does not need steel reinforcement. The sand, cement, and aggregate used in the concrete must be of good quality. The sand and aggregate should be carefully washed to remove any contaminants, such as clay. A high-strength concrete, using higher proportions of cement in the mixture, should be used. Formwork can be made of steel, wood or even mud because it is removed once the concrete has set. Pre-cast concrete sections can also be used, either as main structural members or as 'lost' formwork where the main concrete structure is cast over them.

The space between walls built underground and the undisturbed earth needs to be filled with hard-packed earth to give strength. Masonry is much stronger under compression than under tension, so packing earth tightly around a masonry cylinder actually enhances its strength. The most successful plant designs use partially spherical shapes, because they are inherently stronger, especially under compression, and easier to make leak-tight. The manufacture of a spherical chamber must be subjected to careful quality control because deviations from the correct shape can cause part of the walls to come under tension and then crack.

Cheap building materials that can be used to build low-cost housing are not always suitable for biogas plants. For example, unfired clay will absorb water from the slurry and fail. While wood is used to make barrels for liquid such as beer or water, it is difficult to make it water tight, so is seldom suitable for making biogas plants. Fibrous materials, such as bamboo, have been used to reinforce concrete, but are not suitable unless they are in a matrix of non-porous material. The fibres can absorb water and swell, cracking the concrete. If fibrous materials are used for reinforcement of concrete, they should adhere strongly to the concrete matrix.

Rubber and plastic materials have been used in biogas plants (see Chapter 7), but care must be taken in their use. Many plastic materials are affected by sunlight and degrade with time. Rodents are often attracted to plastics and can cause damage to them. If flexible sheets of material are used to contain the gas, care must be taken to protect them from damage by sharp objects. Birds, animals, and even children are often attracted by these materials and can damage them. Flexible materials must not be brittle – many plastics will crack if they are flexed too many times.

People will not use a biogas plant if it is too difficult to run and maintain. One early design required someone to climb a ladder with a bucket of slurry each day. Not many people are prepared to take on this chore. People are not willing to continue the use of a system that fails frequently and is difficult or expensive to repair. If equipment that needs regular servicing is placed within the digester, the slurry needs to be removed in order to access it, which is difficult. If equipment that is likely to need servicing, such as stirrers and pumps, is used in a digester design, it needs to be placed where it is easily accessible.

In the tropics, the slurry temperature in the plant should remain above 20° C for most of the year. One advantage of an underground plant is that the earth temperature varies very slowly. In colder climates, a biogas plant must be designed so it can be easily insulated and heated to maintain the working temperature. Anaerobic digestion is only slightly exothermic, so the slurry will need to be heated unless the insulation is very effective.

Basic plant designs

While there is a wide range of domestic biogas plant designs in the literature, most can be grouped under three headings: floating drum, fixed-dome, and flexible bag (see Chapter 7).

The designs used in the large programmes in Asia fall under these three headings. Each of these systems includes a limited ability to store gas within the gas plant, usually for a period of half-a-day when the plant is running at its design capacity. Other systems use a fixed digester volume with a secondary gas store. Because this approach involves greater expense, it is not normally used in the programmes that have built large numbers of plants.

While the designs presented in this chapter are based on standard ideas developed over the past 30 years, each group has developed its own variations on the basic design. The programme in Nepal adapted the designs received from India and then China for local conditions. The deep hole required for the standard floating drum design (KVIC design) was impractical in areas with a high water table, so a way of making the same volume of digester in a less deep hole was developed. Other groups have developed a slightly different way to achieve the same result.

Floating drum design

The floating drum design is seen as more expensive than the other design because it uses a steel drum, but it deserves an important place as the design that launched the biogas programme in India. It is still used by some groups in

India, especially for larger volume plants (more than 20 m³ digester capacity). The original idea was developed by Mr J.J. Patel in Bombay, who called it the Gramalaxmi III design (NIIR, 2004). Khadi and Villages Industries Commission (KVIC) has used this design as the basis of its biogas programme since the 1960s. The design was also given extensive publicity in the 1970s by Ram Bux Singh, through reports based on his experiments in the Gobar Gas Research Station in Ajitmal, Uttah Pradesh (Singh, 1971, 1973, 1974).

KVIC design as used in India and Nepal

The digester is a cylindrical pit in the ground, lined with brick masonry (see Figure 6.1). It is built in a very similar way to a dug water well (see Appendix 2), so it relies on expertise that was already available in most villages in India. The floor and walls are made from brick or stone masonry with cement mortar (Bulmer *et al.*, 1985). A pipe is placed in the centre of the pit and a rod with a metal ring is attached to the pipe, so the wall can be built as a cylinder. As the walls are built up within the cylindrical pit, soil is packed behind the bricks.

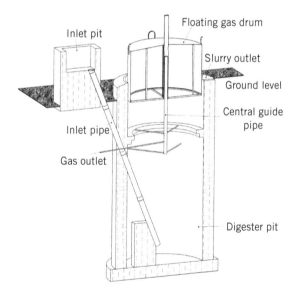

Figure 6.1 Floating drum digester – cross-section view (based on (Fulford, 1988))

The soil around the pit supports the brick walls and allows them to withstand the hydraulic pressure from the slurry inside the plant.

Slurry is fed into the pit through a pipe that finishes near the bottom of the digester pit. The pipe is usually standard drainage pipe, made from spun concrete, ceramic or plastic. A mixing pit is usually built over the mouth of the pipe, so that feed materials, such as animal dung, can be mixed with water before they are allowed to flow into the pit. An outlet is provided at the opposite side of the main pit, often a gap in the top wall through which old slurry can flow when new material is added through the inlet. The effluent slurry (which has minimal smell) is often collected in a shallow pond, where it is allowed to dry in the sun before it is taken for use as compost in the fields.

The slurry in the digester decomposes anaerobically, producing biogas, which is collected in the steel drum floating upside down in the slurry. As the gas collects, the drum floats higher in the liquid. As gas is used from the drum, it sinks back into the slurry. If excess gas is collected, the drum rises to its highest level and the gas bubbles out around the edges. A masonry ledge is provided, so the drum can rest on it when it is at it lowest level, i.e. when it is empty of gas. This ledge also diverts gas coming from the slurry at the edges of the pit into the drum. The inside of the masonry pit is often plastered with cement plaster, especially if the bricks are of poor quality and likely to be porous.

The gas drum, usually made of welded steel, (see Figure 6.2) has a central pipe, which is mounted on a central guide pipe fixed vertically in a frame fitted at the top of the cylindrical pit. These concentric pipes prevent the drum from tipping over and jamming on the walls of the masonry cylinder. They also allow the drum to be rotated around its central axis. Bars fitted inside the drum allow the surface of the slurry to be stirred when the drum is rotated. This helps to break up any scum that may form. The steel drum usually has

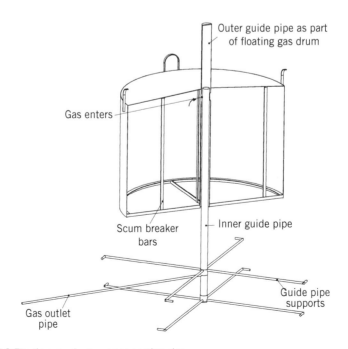

Figure 6.2 Floating gas drum – cross-section view

handles welded to it, which helps the drum to be transported to and installed on the plant, as well as enabling it to be rotated.

Mild steel corrodes very easily, especially in an environment that contains water and mild acids in the slurry and oxygen from the air. Since the steel inside the drum is in an anaerobic atmosphere, corrosion is less of a problem. However, the outside of the drum regularly dips into the slurry and then rises into the air, as it is emptied and filled with gas. A very good paint protection is required to prevent corrosion. The most effective system was found by Development and Consulting Services (DCS) to be a red oxide primer followed by a bituminous top coat (Bulmer *et al.*, 1985). Epoxy or polyester resins will last longer, but are more expensive and difficult to apply. They are also brittle and can flake off, especially when the drum is transported to site.

The basic design for a rural family in India and Nepal has a digester volume of 7.1 m^3 and produced sufficient gas (nominally 2.83 m^3) to cook for a small family from the dung produced by four to six cattle (Fulford, 1988). Designs for larger systems are also available, see Appendix 2 (Devkota, 2001). It is these larger designs that are still being made in India and Nepal. However, there is a limit to the total volume that can be made to this design of plant, as the steel drums must be transported to the site on a truck that can travel on rural roads.

The original design has a gas tap fitted in the top of the gas drum. A flexible plastic pipe connects this tap to a fixed vertical metal pipe, which connects to the gas line that transfers the biogas to where it is needed. The flexible pipe needs to be replaced at least once a year, if not more often, as it tends to become brittle and crack with time. A modification to the basic design (shown in the figure) allows the gas to be removed via the central guide pipe through a gas line that runs underground below the edge of the steel drum (Bulmer *et al.*, 1985).

Another weakness of the original KVIC design is the deep pit that must be dug (3.56 m for the 7 m^3 design). In areas with a high water table, this can be a problem. There have been various designs that seek to provide the same volume with a reduced depth. A wider hole is dug to a reduced depth. The brick walls are tapered inwards (Figure 6.3), so the diameter of the hole at the ledge that supports the drum fits the standard drum for that size of plant (Bulmer *et al.*, 1985). The earth backfill must be carefully rammed into the gap between the brick wall and the undisturbed earth as the wall is built; this gap gets larger as the plant is built. Other groups have adopted alternative approaches. GATE used a semi-spherical wall to adapt the larger base diameter to the diameter at drum level (Sasse, 1988).

One of the reasons for the deep pit used in the standard KVIC design is the idea that biogas digesters should be 'plug flow': the material flows in at one end and out of the other. The DCS tapered-design uses a similar drain pipe to that used for the inlet, to allow the effluent slurry to flow out. If a less deep pit is employed, with material being fed and removed close to the base of the pit, there is a danger of new material 'short-circuiting', leaving the digester before it is fully digested. To avoid this, a central wall is usually built between the inlet and outlet pipes in these wider pits, forcing the fed material to travel to the top of the pit before flowing back down to the outlet pipe near the base.

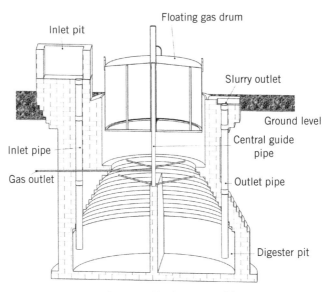

Figure 6.3 Floating drum digester for area with high water table

Further variations on the floating drum design

The standard KVIC design uses an underground pit lined with masonry and a gas drum made from welded steel. In the 1980s, KVIC and other groups experimented with other materials, especially plastic. Large commercial plastic manufacturing plants had been set up in India that made the use of plastics materials much less expensive than when these materials had to be imported (see Chapter 7).

Fixed-dome plants

Traditional Chinese culture encouraged people to be very attracted to the idea of biogas technology, because it encourages the recycling of food and agricultural wastes for use as fertilizer on the land. The first Chinese plants were based on septic tanks (van Buren, 1979), but it was quickly discovered that rectangular shapes tended to crack and cause gas and slurry leaks. Researchers in China came up with the idea of a masonry dome using spherical shapes as far as possible (SADI, 1980). As long as a spherical masonry shape is kept under compression, it remains very strong and unlikely to crack or leak. The best dome shape has an included angle of less than 100° to ensure the masonry remains under compression.

The plants collect gas through the 'displacement principle' (Figure 6.4). As the gas collects in the inside of the dome, it displaces slurry into a reservoir pit. When the gas is used, the slurry returns and refills the space in the digester. There are disadvantages and benefits of this approach. The slurry

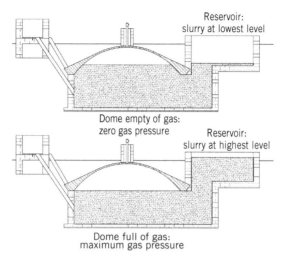

Figure 6.4 Displacement principle for gas storage (based on (Devkota, 2001))

that is displaced is still fermenting, so some biogas is released into the space above the reservoir pit. If this space is open to atmosphere, this biogas is lost and also adds methane (which is a greenhouse gas 20 times more potent than carbon dioxide) to the atmosphere. The gas pressure inside the dome increases as the gas collects, as the hydraulic head between the slurry in the reservoir and the digester increases. This causes increased gas production in the burners that use the gas. However, this higher gas pressure allows smaller pipes to be used for transporting the gas. The gentle flow of slurry back and forth between the digester and reservoir causes mixing of the slurry within the plant, which enhances gas production.

The release of methane into the atmosphere from the reservoirs, though small, is a concern to those who see climate change caused by greenhouse gases as a significant problem. A case can be argued that the amount released is much less than that from the situation when animal dung or food residues are not collected for use in biogas plants and therefore generate methane and other gases as they decompose. The replacement of wood fuel with biogas saves the pollution generated from inefficient combustion of wood, which generates other greenhouse gases. If biogas is used to replace LPG, it is saving fossil carbon from being released into the atmosphere.

The fixed-dome design (see Figure 6.5) formed the basis of the very large biogas programmes in China, the first in the late 1970s and early 1980s, and the more recent programme that is still continuing (NSPRC, 1984). The designs used in both programmes are very similar; the main difference is the higher degree of quality control employed in the later programme, which has allowed the plants built under this programme to be much more reliable.

The fixed-dome design has been adapted by programmes in India and Nepal, although different versions have been made and become popular in various places. Although a range of designs has been used, the three that have

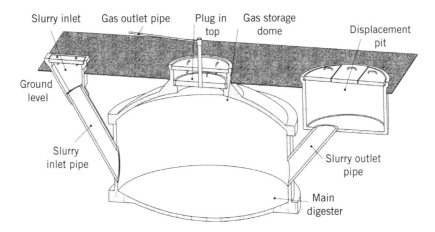

Figure 6.5 Fixed-dome design of biogas plant – cross-section view (based on (NSPRC, 1984))

proved most popular are the Janata and Deenbandhu designs in India, and the GGC design in Nepal. The Janata plant is a fairly close copy of the original Chinese fixed-dome design, while the Deenbandhu is an adaptation of a more recent approach developed in China.

There are two ways to make these designs of biogas plant, either by using bricks and mortar or by using cast concrete. Certain designs lend themselves more to one approach than the other. The Chinese fixed-dome and Janata designs can be made using either bricks or concrete, while the Deenbandhu plant is more usually made of bricks. The GGC design is usually made from concrete in Nepal, especially in the more mountainous regions, where bricks are difficult to obtain, but broken stone and sand are both easy to collect from hillsides and riverbeds.

One major feature that may or may not be included in these designs is a central concrete plug in the roof of the dome that can be removed for access. This concrete plug was an important feature of the design used in the early Chinese programmes; it allowed the plants to be completely emptied every six or twelve months. This fitted in with the idea that the plant was a store of fertilizer for the land as well as a way to generate biogas. However, this concrete plug needs to be carefully sealed against gas leaks, which becomes more and more difficult the more times that it is removed and replaced in the concrete collar at the top of the plant. The usual way to seal this concrete plug is through the use of a layer of wet clay. As long as the clay remains moist, it should remain gas tight, so water is placed in a shallow container made in the top of the plant. A concrete cover is usually placed over this container to reduce loss of water by evaporation.

Designs that do not use a central concrete plug need to provide a larger opening to the slurry reservoir through which a person can enter the plant. This is usually necessary while the plant is being built, as the inside of the dome is plastered to reduce leakage of gas through the porous masonry. This type of plant can be emptied of slurry through the slurry outlet, but this is only required if problems are encountered, such as antibiotics entering the plant and killing the bacteria.

The Chinese fixed-dome or Janata design

As with the floating drum design, the Chinese fixed-drum design uses a masonry-lined underground pit. The pit is shallower and wider than a floating drum design, but does not use a central wall. The movement of slurry in and out of the pit as a result of the displacement principle causes the slurry to be well mixed. These designs use an inverted dome-shaped floor, of a similar shape to the dome that forms the roof. The side walls of the digester can be either straight, forming a cylinder, or slightly inclined outwards, forming a section of a cone. The conical shape is used where the surrounding soil is less stable. Pipes are set either side of the main digester chamber for input and output of slurry. The pipes can be made from spun concrete or plastic. More than one inlet pipe may be used, for example, if a latrine is attached that exits directly into the digester pit.

The domed roof is usually made over formwork to give it the correct shape. Deviations from a spherical shape will cause the dome to be weakened. The formwork can be made from wood or metal. An alternative approach is to fill in the digester pit with mud and shape the top surface to form a mould. The mud must be removed, once the dome has set, so this is a fairly time-consuming option. It is used in places where labour is cheap. The floor, walls, and dome can be made from bricks and mortar, concrete blocks and mortar, or from cast concrete.

If an access hole is installed in the top, the dome is completed with a concrete ring that has one or more ledges made into it. The lower ledge supports a concrete plug that is cast in a separate mould. This concrete plug usually carries a metal pipe that is used to remove the biogas from the dome. The space between the plug and the inside of the concrete ring is filled with wet clay to seal it against gas leaks. A cast concrete lid can be placed on an upper ledge to reduce the rate at which water evaporates from the wet clay.

A slurry-mixing tank is built over the inlet pipe and a slurry reservoir is built over the outlet tank. Earth is piled over the dome to a depth of at least 0.5 m, to provide weight to ensure the masonry remains under compression. Metal or plastic pipe is connected to the gas outlet to transfer the gas to where it is needed.

Deenbandhu design

The Deenbandhu plant (Figure 6.6) uses a similar floor to that of the Janata plant. As the whole plant is usually made from bricks and mortar, the floor is laid with bricks placed into a spherically shaped hole. The top edge of the floor has a ring of bricks placed horizontally to form the foundation for the dome. The key to making this design correctly is a steel pin fitted firmly in the centre of the floor or on a short brick plinth (Figure 6.7). This pin locates

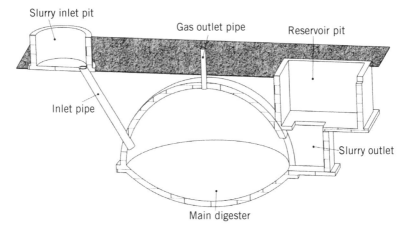

Figure 6.6 Deenbandhu design of biogas plant – cross-section view (based on (VK NARDEP, 1993))

the centre for a piece of wood with a hole or slot in one end that acts as a guide for placing the bricks. It is essential that this centre is solidly fixed while the plant is built, so the dome is of the correct shape. The length of wood has a mark or a transverse piece of wood near the other end to define the radius of the dome.

The dome is built of layers of bricks placed around the circle, each placed so the edge of the brick matches the mark on the guide. The bricks in the lower rings are held in place by the mortar used to lay them. The bricks in the higher rings would tend to fall off until a whole ring is completed, in which case the whole ring locks itself in place. There are two type of simple clamp that are used to hold bricks in place until the ring is complete. The first is a length of wood similar to the guide, with a brick held in a wire or rope noose tied to it. The other end of the length of wood is placed near the centre point, so the weighted length holds a brick in place (Figure 6.7). The other type of clamp is made from stiff wire forming an 'S' shape. A spare brick is placed in one loop if the 'S', while the other is placed over a brick that has been laid in the ring (Figure 6.7).

The top of the dome can be finished off in one of two ways. If a manhole is to be made in the top, a pre-cast concrete ring can be dropped into the last brick ring so it locks the whole structure in place. If a manhole is not used, a wooden or metal form can be used to support the last few bricks that are used to fill in the hole left at the top of the dome. The gas outlet pipe must be fitted at the top of the dome.

The slurry inlet and outlet pipes must be fitted into the right places as the dome is being built. If a top manhole is not used, a larger outlet space is required in the side of the dome, so the builders can get in and out. An inlet pit is built over the inlet pipe and a reservoir built over the outlet pipe or outlet space. Earth is piled over the dome to add weight to ensure the masonry remains under compression.

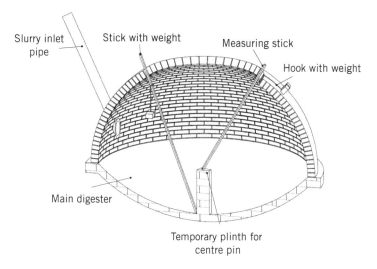

Slurry inlet pipe

Stick with weight

Measuring stick

Hook with weight

Main digester

Temporary plinth for
centre pin

Figure 6.7 Building a Deenbandhu design – cross-section view

GGC design

The original GGC design in Nepal used a domed floor to the digester pit, but experience with the floating drum design suggested that a flat floor works as well. A pit is dug to the correct size and the floor and walls built of brick or stone masonry. Soil is packed around the walls to give them support to withstand the hydraulic pressure within the digester pit. Inlet pipes are fitted as the wall is built and an opening is left opposite the main inlet pipe, which links with the outlet and reservoir chambers, when they are built.

Once the pit is completed, soil is piled back into the hole. A vertical pipe is placed at the centre of the hole and held vertical to act as a centre around which a template can be rotated. The sides of the top of the hole are carefully shaped using the template, to act as an area to take the weight of the dome. The earth in the hole is piled up to form a dome shape, which is also shaped using the template. The dome is then cast over the earth mould using concrete. The edges are made thicker to act as a foundation for the rest of the dome. Two 6 mm-diameter steel reinforcement-rods are placed in the part of the edge that crosses the opening in the wall. The whole dome should be cast in one session, because joins between parts cast at different times can result in cracks that leak gas. When the concrete is set, the earth is dug out of the hole again, through the opening in the side of the pit. The outlet and reservoir chambers are built around this opening.

The gas outlet pipe is fitted over the central guide pipe when the dome is being cast. The guide pipe can be removed once the dome is set. Often a masonry structure (the turret) is built around the gas outlet pipe to support it (Figure 6.8).

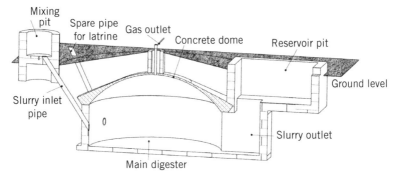

Figure 6.8 GGC design of biogas plant – cross-section view (based on (Devkota, 2001))

Features common to dome designs

Sealing the dome

The inside of any of these domes must be plastered with a suitable plaster to make it gas tight. There are various options. The Chinese manuals (NSPRC, 1984) suggest two or three layers of cement plaster (with 1 part cement to 2.5 parts sand), each painted carefully with cement paste after it has dried. Waterproof paints can be used to fill the pores of the plaster and prevent it from being porous to gas. Water-based paints should be used, as two-part mixes and those that use other solvents give off fumes that make them dangerous to apply in the confined space of an underground dome.

There is a wide range of commercial plastic emulsions that polymerize as the paint dries, giving an effective impervious layer. The plastic in these emulsions can be PVC, epoxy, acrylic, or silicone based. People in China have used PVC emulsions that are often used to give a waterproof wall coating in bathrooms and kitchens (Tang *et al.*, 1985). In Nepal, an acrylic plastic emulsion has proved very effective (Bulmer *et al.*, 1985). In India, epoxy-based emulsions have been successfully used to repair leaky domes (VK NARDEP, 1993). The concern over pollution from solvent-based paints in recent years had led to the development of a wide range of water-based paints that should seal concrete very effectively. There are new siloxanic paints that incorporate silicone with plastics that should give even better performance.

If a water-based paint is used, it can be mixed with cement and sand (1:2) and water to form a plaster layer. The plaster layer needs to be completely applied in one session and then allowed to dry. Once this plaster has dried, it can then be coated with a paste made from cement powder, paint, and water.

The reservoir pit

The reservoir pit is used to store the slurry when it is displaced from the main digester pit; it must store an equivalent volume of slurry to the amount of gas which displaces it. A typical storage volume is 50 per cent to 60 per cent of the daily gas production. If the gas is used for cooking, it must collect overnight, so it can be used during the day. Reservoir pits can be cylindrical or rectangular in shape, because they do not need to be gas tight. The depth of liquid in the reservoir is less than in the main pit, so the risk of leakage of slurry from cracks in the corners is less. If space is limited, the reservoir pit can be partially built over the main digester pit, so part of the main dome can form part of the floor.

If the dome plant has a central manhole, the reservoir pit is usually connected to the main digester pit using a length of drainpipe, similar to that used for the slurry inlet. If there is not a central manhole, there must be an opening in the side of the main digester pit, large enough for a man to enter through (at least 600 mm × 600 mm). This opening needs to connect to an access chamber that leads up into the base of the reservoir pit, which is also large enough for a man to climb down with the help of a ladder.

The level of the floor of the reservoir pit is important. Ideally, the slurry should just reach this level when the dome is empty of gas. The level of slurry inside the dome should be well below the top of the dome (about one quarter of the radius), so that slurry cannot enter the gas outlet pipe. An outlet is made in the side of the reservoir pit, at a height so that the slurry flows out from it when the digester pit is almost full of gas. The volume of slurry contained in the reservoir pit, from the level at the floor to that at the slurry outlet, should be smaller than the volume of the gas storage dome, from the top level to where the gas would begin to bubble out of the dome through the slurry outlet.

The head of slurry between the lowest level inside the gas dome and the level of the slurry outlet determines the maximum gas pressure in the plant. A higher gas pressure will encourage more gas leaks, so this value should be kept below 1 m for smaller plants and 1.8 m for larger plants.

Pit covers

A reservoir pit is an open tank with its edge close to ground level, so there is a danger that people and animals can fall into it. It is usually covered with removable flat covers, made so that one person can lift each cover.

Covers are usually made of concrete reinforced by steel rods, usually about 50 mm thick. Covers are usually made in shaped moulds, which can be made from mud, steel or wood. Steel bars are usually bent into handles, so the covers can easily be lifted. The handles can either be fixed in the concrete or placed in larger pre-made holes, so they drop down when they are not needed. Some reservoirs have fixed roofs, with removable covers only over the main access hole.

One type of cover that needs to be very carefully made is the plug for the manhole in the top of the dome, if that type of design is used. The outside diameter of this plug must be about 5 mm smaller than the hole it fills, so that clay can be pushed between the plug and surround to seal the opening. If the hole has vertical sides, the plug is often tapered towards the centre, giving a wedge shaped gap into which clay can be pressed. An alternative approach is to make both the hole and the plug tapered downwards, so the plug locks into the hole. The clay must be shaped around the plug as it is inserted. The disadvantage of this approach is that every time the plug is removed, the sides can wear, so the gap becomes larger.

A more recent approach for a Deenbandhu plant is to use a spherical section to cover a circular reservoir pit. If masons are skilled in using the Deenbandhu approach to making a dome from circles of bricks, a domed cover is much cheaper than using a flat concrete cover reinforced by steel rods.

Slurry inlet pit

These underground biogas plant designs are hydraulic, allowing slurry to flow in and out of the digester pit. A slurry of fine organic particles in water will flow freely, as long as the total solid content is 12 per cent or less (water content 88 per cent or more). Most animal dung has a total solid content that is higher (16 to 20 per cent for cattle dung), so water must be added to allow the slurry to flow. If too much water is added and the total solid content goes below about 5 per cent, the solids and liquids can separate within the digester and the process becomes less efficient.

The inlet pit is often used to mix the water and organic materials to give a uniform slurry with no lumps, that will flow freely. The mixing can also be done in a separate vessel and the slurry poured into the inlet pipe. If this is the procedure to be adopted, then the inlet pit can be funnel shaped. If an inlet-mixing pit is made, it must be of a volume that is larger than the daily feed volume of slurry. The shape can be either rectangular or cylindrical.

The bottom of the inlet pit is often sloped downwards away from the entrance to the pipe leading to the digester pit. As the slurry is mixed, inert materials, such as sand, will settle to the bottom of the slope and not enter the inlet pipe. Once the mixed slurry has entered into the digester pit, the inert material can be scooped out and discarded.

A plug must be provided, so that the slurry remains in the pit until it is fully mixed. A plug can be made from wood or plastic, tapered so it fits into the end of the inlet pipe. It can have plastic or rubber wrapped around it, to give a better seal. It can also have a handle attached to the top, so it can be pulled out to allow the mixed slurry to enter the digester pit. Another approach is to use a removable gate between the mixing pit and the entrance to the inlet pipe that is placed just outside the mixing pit. The gate, which is usually made of wood, is lifted to allow the slurry into the plant.

In India and other South Asian countries, the mixing of dung and water is often done by hand. Mixing machines have been designed and used in the Nepal biogas programme. The first mixing machine was based on a design used to mix paper pulp in India and used a set of paddles which rotate on a horizontal axis, driven by a handle (see Chapter 8). Since this splashed slurry over the operator, it was not very popular. A better model uses blades mounted on a vertical rod, driven by a handle, which looks similar to a large food-mixing machine (Devkota, 2001). The blades are mounted in a cylindrical inlet pit made from masonry (see Chapter 8). However, the steel blades do rust away after a few years and need to be replaced.

Biogas plant designs for low temperatures

The large biogas programmes (in China, India, Nepal, and other places in Asia) have proved effective where the local climate provides ambient temperatures that are normally above 28 °C. The designs that are available are not really suitable for being insulated and heated. In colder areas of countries that are running these programmes (such as at higher altitudes), people have devised ways in which the biogas plants can be run (Balasubramaniyam et al., 2008).

A biogas plant is only mildly exothermic: it does generate energy in the form of heat, but the amount is low (see Appendix 1). The amount of heat generated is insufficient to heat a large volume of slurry by more than a few degrees. If the slurry is close to the operating temperature (around 35 °C), the microbial population appears to be able to control its temperature over a small range. Mechanisms are needed that reduce the amount of heat lost from a plant to a very low level, by providing insulation. Also a system is required that can provide heat to the digester to raise its temperature to a value that allows the microbes to function (Bulmer et al., 1985). The provision of external heat is expensive.

Insulating biogas plants

Floating drum plants are difficult to insulate. One way is to build a greenhouse around the drum (NIIR, 2004), using a frame to support transparent plastic or glass panels, but this is expensive. Most types of low-cost transparent plastic have a short lifetime when exposed to sunlight. The greenhouse allows solar energy to heat the drum above the slurry, while reducing heat losses. The heat from the drum does not transfer easily to the slurry. The drum is full of gas and heat cannot transfer downwards by convection in the gas.

Fixed-dome plants are less affected by cold air temperatures, as the slurry takes the temperature of the surrounding soil. The greatest heat loss is through the dome, which is closest to the soil surface. These plants can be insulated by building a stack of straw above the plant. An alternative is to build a compost heap on the soil over the dome because the aerobic compost process generates heat (Bulmer et al., 1985).

Plastic-bag plants (see Chapter 7) have been effectively insulated by placing insulating material over the bag, or by enclosing the plant in a greenhouse (Perrigault et al., 2012). Losses from the sides of the bag are reduced by using straw insulation and thick adobe walls for the trench in which the bag is held.

Some designs of plastic-drum plants are above ground and smaller, if used with food residues, so can be insulated by wrapping them in an insulation blanket. Specially made blankets are expensive, but low-cost ones can be made from plastic bags stuffed with straw.

Heating biogas plants

The use of an external source of energy to heat a biogas plant is usually uneconomic, as the cost will be greater than the value of extra energy that the plant produces. Some of the biogas generated can be used in a boiler to generate hot water, which can be used to heat the plant. Such a system adds complexity to the design and reduces the amount of gas that can be used for other purposes. Most biogas plants are underground, so hot water must be circulated with a pump, which requires a supply of electricity.

The use of solar heat reduces the running costs for generating biogas, but most designs of solar heater add complexity and greatly increase the capital cost of the plant. As most biogas plants are underground, an electric pump is required to pump solar heated hot water through a heat exchanger in the slurry in the main digester. The pump can be driven by electricity from solar photo voltaic panels, but this increases the capital cost even further.

Heating input material

If it is possible to insulate a biogas plant and to heat the slurry to an adequate working temperature, heat can be gained or lost when new feed material is added to the plant. If the plant has a retention time of 30 days, 1/30 of the total volume of the slurry is added each day. It is often easier to heat new slurry before it is added to the plant than heat the slurry that is inside the plant.

A simple solution to heating the input slurry is to mix it each morning and place a frame covered with transparent plastic over the mixing pit, so the slurry is solar heated during the hot part of the day (Bulmer et al., 1985). If the warmed slurry is allowed to enter the pit later in the day, the temperature of the slurry in the main digester tank is slightly increased.

References

Ashden (2007) 'Biotech: Management of domestic and municipal waste at source produces biogas for cooking and electricity generation', [website - case study] Ashden <www.ashden.org/winners/biotech> [accessed 21 July 2014].

Ashden (2005) 'Kigali Institute of Science, Technology and Management (KIST), Rwanda', [website - case study] Ashden <www.ashden.org/winners/kist05> [accessed 21 July 2014].

Balasubramaniyam, U., Zisengwe, L.S., Meriggi, N. and Buysman, E. (2008) *Biogas Production in Climates with Long Cold Winters*, Wageningen, the Netherlands: Wageningen University. <http://www.susana.org/docs_ccbk/

susana_download/2-855-new-study-prepared-for-wecfbiogas-production-in-climates-with-long-cold-winters.pdf> [accessed 22 July 2014].

Bulmer, A., Finlay, J., Fulford, D. and Wong, M. (1985) *Biogas: Challanges and experience in Nepal*, Kathmandu, Nepal: United Mission to Nepal. <www.kingdombio.com/Biogas-vol-I.pdf> [accessed 17 July 2014].

van Buren, A. (1979) *A Chinese Biogas Manual*, London: Practical Action Publications. (Intermediate Techonology Publicatons) <http://developmentbookshop.com/chinesebiogas-manual-pb> [accessed 22 July 2014].

Devkota, G.P. (2001) *Biogas Technology in Nepal: A sustainable source of energy for rural people*, Kathmandu: Mrs. Bindu Devkota, Maipee.

Fulford, D. (1988) *Running a Biogas Programme: A handbook*, London: Practical Action Publications (Intermediate Technology Publications). <http://developmentbookshop.com/runninga-biogas-programme-pb> [accessed 22 July 2014].

Mark, R. and Hutchinson, P. (1986) 'On the Structure of the Roman Pantheon', *Art Bulletin* 68: 24 <www.jstor.org/stable/3050861> [accessed 22 July 2014].

NIIR, B. (2004) *Handbook on Biogas and its Applications*, Delhi: National Institute of Industrial Research. <www.niir.org/books/book/handbook-on-bio-gas-its-applications/isbn-8186623825/zb,,72,a,5,0,a/index.html> [accessed 21 July 2014].

NSPRC, N.S. of the P.R. of C. (1984) *The Collection of Designs for Household Hydraulic Biogas Digesters in Rural Areas*, China State Bureau of Standardization.

Perrigault, T., Weatherford, V., Martí-Herrero, J. and Poggio, D. (2012) 'Towards thermal design optimization of tubular digesters in cold climates: A heat transfer model', *Bioresource Technology* 124: 259–268 <http://dx.doi.org/10.1016/j.biortech.2012.08.019> [accessed 21 July 2014].

Rajendran, K., Aslanzadeh, S. and Taherzadeh, M.J. (2012) 'Household Biogas Digesters - A Review', *Energies* 5: 2911–2942 < http://dx.doi.org/doi:10.3390/en5082911> [accessed 21 July 2014].

SADI (1980) 'Biogas in China', in *Technical Consultations Among Developing Countries on Large-scale Biogas Technology in China*, pp. 67–117, Beijing, China: United Nations Industrial Development Organization.. <www.worldcat.org/title/technical-consultations-among-developing-countries-on-large-scale-biogas-technology-in-china-beijing-china-4-19-july-1980/oclc/226046783> [accessed 21 July 2014].

Sasse, L. (1988) *Biogas Plants*, GATE, GTZ Eschborn, Germany. <http://www.gate-international.org/documents/publications/webdocs/pdfs/g34bie.pdf> [accessed 21 July 2014].

Sharma, S. (2013) 'Visiting Golghar in Patna', [website - tourist] Golghar in Patna <http://www.shalusharma.com/golghar-in-patna/> [accessed 21 July 2014].

Singh, R.B. (1974) *Biogas Plant: Generating methane from organic wastes*, Gobar Gas Research Station, Ajitmal, Etawah, India.

Singh, R.B. (1973) *Bio-Gas Plant: Generating Methane from Organic Wastes & Designs with Specifications*, Gobar Gas Research Station, Ajitmal, Etawah, India.

Singh, R.B. (1971) *Some Experiments with Bio-gas*, Gobar Gas Research Station, Ajitmal, Etawah, India.

Tang, Z.G., Xie, X.U. and Wu, D.C. (1985) 'Study on Polymer Seal Paint of Concrete Biogas Digester', in *Proceedings of the Fourth Annual Symposium on Anaerobic Digestion*, Guangzhua, China: China State Biogas Association.

Tuladhar, S.M., Ban, B., Shrestha, B.K., Sharma, C.K. and Ranjit, P. (1985) *Biogas Plants in Nepal: An evaluative study*, New Era for UNICEF, Kathmandu, Nepal. <www.snvworld.org/en/download/publications/biogas_plants_in_nepal_an_evaluative_study_1985.pdf> [accessed 21 July 2014].

VK NARDEP (1993) *Biogas: A manual on repair and maintenance*, Kanyakumari: Vivekananda Kendra, Natural Resources Development Project. <http://www.vknardep.org/publications/english-books/197-biogas> [accessed 21 July 2014].

CHAPTER 7
Plastic biogas-plant designs

Abstract

There are two main types of plastic biogas plant: those that use flexible plastic, and those that use rigid plastic to make plants that are similar to floating drum or underground plants. Flexible plastic systems include 'bag' digesters, where the whole plant is made from plastic and 'membrane' digesters where the plastic membrane acts as a gasholder over a pit made from masonry or other materials. Floating drum designs include systems in which the tank and gas holder are both made of plastic (Martin, 2008; Roos et al., 2004) or ones in which only the gasholder is of plastic and the digester tank is made from masonry or other materials. Underground plants are also available in which the digester pit and reservoir are both made of plastic.

Keywords: biogas, anaerobic digestion, plastic plant designs

Overview of plastic biogas plants

In addition to the designs of plants that have been built in numbers amounting to thousands or even millions, other designs have been used by programmes that have built several hundred or even several thousands. The most notable of these designs are the bag digester, which uses a plastic tube as the main digester, and the membrane reactor, in which the digester pit is covered by a flexible membrane to retain the gas. About 22,000 bag digesters were built in Vietnam at the end of the 1990s because they were seen as very cheap and easy to install. Groups in Ethiopia, Tanzania, Cambodia, and Bangladesh have set up biogas programmes using this design. The growing biogas programme in Latin America, in Columbia, Bolivia, Mexico, and elsewhere, has been based on the use of bag digesters.

A similar approach (although for much larger systems) was adopted by the AgSTAR (Martin, 2008; Roos *et al.*, 2004) programme in the USA, which encouraged the installation of 151 biogas systems on farms in the USA, many of which were slurry storage-lagoons covered by a flexible membrane. These membranes are made of more expensive materials, such as high-density polyethylene (HDPE) or polypropylene. Similar membrane-covered lagoons have been used in other countries, for example in Australia and South Africa.

The use of covered slurry lagoons was popular in the USA until 2009 (EPA, 2013), when other designs took over, such as mixed digesters, using metal tanks, and 'plug flow' digesters, using concrete troughs. These digesters are often covered by a flexible sheet that allows gas to collect under it.

http://dx.doi.org/10.3362/9781780448497.007

The development of a particular type of PVC called 'Red Mud Plastic' in Taiwan (Hao *et al.*, 1979) encouraged the wider use of the bag digester design. Red Mud Plastic uses a PVC base with the waste mud from bauxite processing as filler. The plastic is very flexible and is much less brittle than many other forms of PVC. The red mud helps protect the polymer from the ultraviolet components of sunlight, which causes many types of plastic material to degrade. Bag digesters made from this material became popular in mainland China and other parts of the world for a period of time. However, this type of plastic has become less easily available and more expensive. The Taiwanese manufacturers now appear to be using the red mud filler to make a more rigid PVC material that can be used as a roofing material.

More rigid plastic materials, such as HDPE and fibre (glass) reinforced plastic (FRP), can be used to replace steel in floating drum digesters. Underground plants are also made from plastic.

Vietnamese bag digester

Descriptions of the way that bag digesters are made have been published in various papers (Furze, 2002; Lüer, 2010; Rodriguez and Preston, 1998). Vietnamese bag digesters are made from polyethylene tubing, which is widely available in most parts of the world. The diameter of the tubing varies depending on the machines installed by the manufacturers, but sizes range from 800 mm to 2 m. The tubing is usually laid flat and then made into a roll. The width of a roll is therefore half the circumference of the tube (i.e. 1.26 m to 3.14 m wide). The edges of the roll are very vulnerable to damage, which is revealed as small holes along a length of the tubing, so the roll needs to be handled very carefully. The thickness of the plastic is usually between 0.2 mm to 0.25 mm (200 µm to 250 µm).

Two or three lengths of plastic are usually used, one inside the other, so the tube has two or three layers. The outside layer is usually seen as sacrificial; if it is damaged by sunlight or abrasion, the inner layers are protected and maintain the gas tight seal. The length of plastic tubing required depends on the diameter of the bag and the total volume of digester required. Two or three pieces of plastic tubing are pulled off the roll and cut to the length required. They are then put together to form a tube made of two or three layers.

A trench is dug in the ground to take the shape of the bag. The trench needs to be carefully shaped, with appropriate length and width, so that the bag will fit tightly into the trench when it is full of slurry. The floor of the trench should slope downwards at a slope of 2.5 per cent from the inlet to the outlet. The ends of the trench will have smaller trenches that slope to the surface to support the slurry inlet and outlet pipes. The soil dug from the trench needs to be placed well away from it, otherwise rain can wash it back in at a later time.

The gas outlet pipe is fitted to the plastic bag by cutting a small hole and using threaded PVC pipe fittings (usually 12.5 mm diameter) with rubber and plastic washers to make a gas-tight seal through the hole. Slurry inlet and outlet pipes can be made of plastic or ceramics and are usually around 150 mm in diameter and 1000 mm long. Each end of the plastic tube is carefully folded around one of these tubes and held tightly in place with rubber straps cut from old inner tubes.

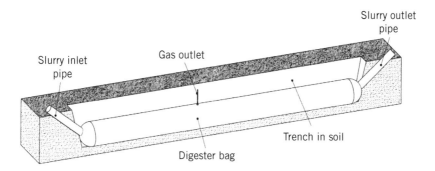

Figure 7.1 Bag digester in trench, (based on Lüer, 2010)

The pipes can be sealed with plastic and rubber bands and the tube blown up with air to check for leaks. The bag should be carried to the trench when full of air so that it can be properly located in the trench. The gas outlet fitting can be connected to a short length of rigid PVC pipe with glue, which can be used to connect to the flexible pipe that conveys the gas to the kitchen. The gas pipe should be supported, so it does not put strain on the bag when it flexes.

The slurry inlet pipe can be connected to a mixing pit made of bricks, with a channel leading from the pigsty through which dung can be washed. The slurry outlet pipe is led into a pond in which the effluent can be stored and dried before being taken to the fields.

Using a bag digester

Polythene plastic materials are badly affected by the ultraviolet components of sunlight so will degrade after a few months. The bag can be placed under a roof or cover made from grass matting to help protect it. Walls of grass matting are also helpful as plastic often attracts rodents and scavenging birds, which have learnt that food is often found in plastic bags.

A bag digester is 'plug flow' – the different biogas digestion processes happen in different places along the tube. Microbes cannot travel easily against the main flow, therefore it is important to mix some of the effluent (10 to 20 per cent) back in with the new feed, especially if pig dung or food wastes are used as the main feed. When the digester is first started, the new feed can be mixed with slurry from a working plant using cattle dung, in order to start the digestion process more quickly.

The top of the bag can be used for gas storage, but as the total volume is fairly small, a second plastic bag can be used to collect and store the gas. The gas pressure within the bag is lower than that from other designs of plant, as the hydraulic head between the liquid levels in the feed and outlet pipes, and the level inside the bag, is small (a few mm). If a second plastic bag is used to store the gas, it is possible to close the pipe leading from the plant and increase the pressure by placing weights on this bag or rubber bands around it, so that gas burners can operate more effectively.

The polythene material has a fairly short life, even when it carefully protected. The main advantage of this design is that it is very cheap.

Latin American bag and tent digesters

There are projects in several Latin American countries that are using bag digesters made from more expensive, but better quality plastic. These bags are placed in trenches lined with concrete or brick (Martí-Herrero and Cipriano, 2012, 2014), so they have better protection than those in simple earth-lined pits. Insulation can be placed around the bag to reduce heat loss enabling the system to be used in colder mountainous regions such as Bolivia (Perrigault *et al.*, 2012). Bag digesters are usually placed under a cover, to protect them from sunlight (Perez Bort, 2011). An adaptation for cold climates is to place the digester under a greenhouse. Walls, made from adobe are made either side of the digester trench, with a higher wall on the north side. Transparent plastic sheets are used to cover the span between the walls, so that the space between can be solar heated (Lüer, 2010).

The material for making the bags can be PVC, polypropylene or high-density polyethylene. The plastic 'sausages' are made from welded plastic sheets to give a suitable shape, with a larger diameter central section, tapering to smaller diameters where the inlet and outlet pipes are connected. The use of PVC for both the pipe work and the sheet material allows for the whole plant to be solvent welded together (Viquez Arias, 2009).

Plants of larger volume can be made as a pit lined with concrete blocks and covered with a plastic sheet (Carmona, 2007) in the form of a tent anchored at the bottom. The edges of the sheet are folded back and glued to the main sheet to form long pockets through which plastic pipe can be threaded. The pipe in these pockets is held in place against the walls of the pit with plastic hooks, to ensure the edges of the plastic remain under the surface of the slurry. The slurry inlet and outlet pipes are placed so they enter the walls at opposite ends of the main pit.

The gas outlet from this type of digesters needs to be carefully made, because the connection moves as the plastic bag or sheet fills and empties of gas. A hole needs to be cut in the plastic sheet, so that a plastic-pipe tank-connector fitting can be passed through it. The area around the hole can be reinforced with extra layers of sheet glued to onto the main plastic sheet. Rubber washers can be used between the flat faces of the pipe fitting and the plastic sheet. If both the sheet and the fitting are made from PVC, glue can be used to strengthen the joint and further seal it. The gas outlet pipe needs to be supported in a way that reduces the strain on this fitting, while allowing it to move as the plant fills and empties of gas.

Covered lagoons

The use of large lagoons to hold slurry from large animal operations is fairly common in the USA and some other countries, such as Thailand (Thiengburanathum, 2010). The need to cover these lagoons was initially inspired by a desire to reduce the odour from them. An effective cover results in anaerobic conditions within the lagoon and the generation of biogas, which must then be collected and used. The anaerobic process within the lagoon also reduces the smell, so spent slurry can be stored in a second, open, lagoon until it can be used as a fertilizer on crops.

The lagoon covers need to be made of a material that is strong, impervious to gas and which can be fabricated into the correct shape in a way that does not introduce holes and therefore leaks. The fabrication technique is usually by welding. Suitable materials are HDPE, polypropylene, and specialist fibre-reinforced flexible-polymer geo-membranes.

The membranes either cover the whole lagoon or are modular in nature and each one only covers part of the slurry area. The complete membrane covers are usually attached to the banks of the lagoon by burying the edges in trenches. Modular covers can be held in place by ropes and need to have sufficient buoyancy to float on the surface of the slurry. A membrane cover must also prevent rain entering the slurry pit and diluting the slurry.

The cost of the material depends on its quality and potential life. Cheap lagoon covers made of thinner material will need replacing more often than ones made from more expensive and thicker materials. One typical lagoon that AgSTAR has studied (Martin, 2008) designed for use with 1,500 to 1,600 cattle was 180 m long by 60 m wide and 6.4 m deep. The total membrane area was therefore 10,800 m^2. The biogas generated fuelled a 180 kW electric generator.

One weakness of a covered lagoon is that there is no temperature control of the slurry. Biogas will be produced during periods when the ambient temperature is above about 25 °C, but quickly ceases below this temperature. Full biogas production requires an ambient temperature of 35 °C.

Floating plastic-tank designs

The initial success of the floating steel-drum KVIC design in India encouraged various groups to consider making such drums of plastic. Domestic biogas systems using food wastes as a feedstock require much smaller plant sizes that can be made more economically from plastic, as the plant sizes are much smaller.

One approach has been pioneered by ARTI, working in Pune, India (Ashden, 2006a; Vögeli et al., 2009). The design uses cylindrical water tanks made from HDPE. Plastic water tanks are mass-produced in very large numbers in many countries, so have a low unit cost. Two water tanks of different sizes are used, which are cut and hand welded to form two containers that telescope together (Figure 7.2). The lower (larger) container acts as the main digester, while the upper inverted container is the gasholder. The initial design used a lower container with a capacity of 1 m^3, so the gasholder was made from a 0.75 m^3 water tank. Larger systems can be made using water tanks with larger volumes. However, the unit cost rises rapidly with size, as the demand for large water tanks is much lower than for the smaller domestic sizes.

Components, such as tanks, made from HDPE can be made at low cost as long as there is a very large demand for them. The metal moulds from which the shapes are made are very expensive, so large numbers of units must be made from these moulds in order for the cost of each item to be low. The demand for small-scale biogas plants in India could be sufficient for such moulds to be made, as long as a standard design can be developed that would have a large potential market. This is an approach being adopted by VK NARDEP in South India (Ashden, 2006b). The NGO is testing out a range of designs made of other materials, such as FRP and steel, so that a mould could be made based on the design that proves to be the best. The company is also researching the potential market for such biogas plants to find out if there might be a demand for thousands of such units.

One of these designs has been produced by a plastic-moulding company (Sintex, 2013) is already making both floating drum and fixed-dome biogas plants from moulded HDPE (Chittora, 2008). These plants are now being sold by international resellers, although it is not clear how many of these plants are being installed in India. Sintex Ltd is not on the list of approved suppliers defined by MNRE.

FRP tanks can be made using much cheaper moulds. Biotech Ltd, an Indian company (Ashden, 2007a; Estoppey, 2010), has been manufacturing both gas holders and complete biogas plants from this material. The complete FRP units have a main digester tank, with an inverted gasholder tank that telescopes

Figure 7.2 ARTI floating drum plastic biogas unit from HPDE water tanks

into it. The main digester can also be made from pre-cast reinforced concrete rings, but these are usually placed underground. The overall cost of a FRP plant is higher than one made from HDPE, but there is a steady demand for these units in south-west India. Biotech Ltd now also makes similar plants from moulded HDPE because this market is very large (Figure 7.3).

The FRP composite is laid-up on a metal mould and allowed to cure. The structure usually includes a metal frame, especially around the open rim of each tank, so it keeps its shape. The composite drum is much lighter and easier to transport than one made purely from steel. FRP drums are usually limited to smaller sizes of biogas plant, as larger-sized drums tend to crack when transported. A typical working volume of such a plant is between 1 m^3 and 2 m^3.

These plastic digesters work well in South India, where the ambient temperature is normally high enough to allow effective digestion throughout the year. Such plastic units could be insulated to retain heat in colder climates. The amount of material required to make a double-skinned plant would be greater resulting in a higher overall cost.

FRP and other plastic drums are much lighter than steel ones, so weight has to be added to give internal pressure within the gas volume. Weights can consist of a concrete block placed in a frame fitted to the top of the drum. It is also possible to make a container in the top of the drum that can be filled with sand. Alternatively, pipes filled with sand or gravel can be added than run from the top to the bottom inside the drum. This approach has the advantage that the centre of gravity of the drum is lowered, making it more stable and less likely to tip over.

Using food wastes from an individual household in India allows the owner of a plant to save 25 per cent of the LPG they would have otherwise used.

Figure 7.3 Biotech floating drum plastic biogas unit made from HPDE

Many plant owners seek outside sources of food waste, for example from local small food processing industries, so that they can use biogas to provide more of their cooking needs.

Underground plastic biogas plants

The popularity of underground biogas plants has led to the development of such plants made from plastic. Several commercial and semi-commercial projects in China now manufacture different types of underground plants made from FRP and other types of plastic (Cheng *et al.*, 2013). These plants are also being exported to places such as Bangladesh and Vietnam. Designs include masonry pits with a fixed plastic-dome cover and whole spherical units made from plastic.

SimGas BV has developed an underground design of plastic biogas plant made as a horizontal cylinder from prefabricated sections (Simgas, 2014). It is a displacement digester, with the plastic reservoir mounted on top of the cylinder. It is being installed in Tanzania and Kenya.

Biotech Ltd has been involved in biogas plants that use food waste and sewage from institutions and markets (Ashden, 2007b). A more recent innovation is such a plant made from FRP. It is a two-stage digester, with cylindrical pre-digester tanks mounted above main digester tanks, so that the recycled liquor flows through the system by gravity. The main digester tanks are partially buried cylinders placed at a slope to the vertical (Figure 7.4). The unit is modular and can be made of a number of pre-digester and main digester units.

The Biotech system does not use the displacement principle: the main digester tanks are filled with slurry and the gas is collected in a flexible plastic

Figure 7.4 Biotech food waste plastic biogas unit made from FRP

bag, protected by a lightweight FRP tank. The system is seen as more expensive than some other designs that could process the same amount of material.

References

Ashden (2006a) 'Appropriate Rural Technology Institute: Compact digester for producing biogas from food waste', [website - case study] Ashden < www.ashden.org/winners/arti06> [accessed: 21 July 2014].

Ashden (2007) 'Biotech: Management of domestic and municipal waste at source produces biogas for cooking and electricity generation', [website - case study] Ashden <www.ashden.org/winners/biotech> [accessed: 21 July 2014].

Ashden (2006b) 'VK-NARDEP, India: Multiple benefits from biogas', [web site - case study] Ashden <www.ashden.org/winners/vknardep> [accessed: 21 July 2014].

Carmona, T. (2007) *Understanding the Basics of a Biodigester,* [web page] Biodigester Design & Construction Rural Costa Rica, San Jose, Costa Rica. <www.ruralcostarica.com/biodigester.html> [accessed: 21 July 2014].

Cheng, S., Li, Z., Mang, H.-P. and Huba, E.-M. (2013) 'A review of prefabricated biogas digesters in China', *Renewable and Sustainable Energy Reviews* 28: 738–748 <http://dx.doi.org/10.1016/j.rser.2013.08.030> [accessed: 21 July 2014].

Chittora, M. (2008) 'Sintex (India) Introduces Ready-to-Use Biogas Plant', *Hedon* Article 1522 <www.hedon.info/article1522> [accessed: 21 July 2014].

EPA (2013) *AgStar Accomplishments,* USA: Environmental Protection Agency. <www.epa.gov/agstar/documents/2012accomplishments.pdf> [accessed: 21 July 2014].

Estoppey, N. (2010) *Evaluation of small-scale biogas systems for the treatment of faeces and kitchen waste: Case study Kochi, South India,* Dübendorf, Switzerland: Swiss Federal Institute of Aquatic Science and Technology (Eawag). < www.eawag.ch/forschung/sandec/publikationen/swm/dl/Estoppey_2010.pdf> [accessed: 21 July 2014].

Furze, J. (2002) *Tubular Plastic Bio-Digesters in Tanzania, Vietnam, Zimbabwe & China,* Denmark: University of Aarhus.

Hao, P.L.C., Hsu, W.W. and Tang, H.S. (1979) 'PVC red-mud compositions' Patent <www.google.com/patents/US4161465> [accessed: 21 July 2014].

Lüer, M. (2010) *Installation Manual for Low-Cost Polyethylene Tube Digesters,* Bolivia: GTZ. <https://energypedia.info/images/1/19/Low_cost_polyethy-lene_tube_installation.pdf> [accessed: 21 July 2014].

Martí-Herrero, J. and Cipriano, J. (2012) 'Design methodology for low cost tubular digesters', *Bioresource Technology* 108: 21–27 <http://dx.doi.org/10.1016/j.biortech.2011.12.117> [accessed: 21 July 2014].

Martí-Herrero, J.; Chipana, M.; Cuevas, C.; Paco, G.; Serrano, V.; Zymla, B.; Heising, K.; Sologuren, J.; Gamarra, A. (2014) 'Low cost tubular digesters as appropriate technology for widespread application: Results and lessons learned from Bolivia' *Renewable Energy* 71: p156–165 <http://dx.doi.org/10.1016/j.renene.2014.05.036> [accessed 21 July 2014].

Martin, J.H. (2008) *An Evaluation of a Covered Anaerobic Lagoon for Flushed Dairy Cattle Manure Stabilization and Biogas Production,* AgSTAR Environmental

Protection Agency, USA. Washington DC<www.epa.gov/agstar/documents/flushed_dairy_cattle.pdf> [accessed: 21 July 2014].

Perez Bort, I. (2011) *Optimization of the biogas digesters: Single family homes in the rural area of Andean Peru*. MSc Thesis. Barcelona: Universitat Politecnica de Catalunya.

Perrigault, T., Weatherford, V., Martí-Herrero, J. and Poggio, D. (2012) 'Towards thermal design optimization of tubular digesters in cold climates: A heat transfer model', *Bioresource Technology* 124: 259–268 <http://dx.doi.org/10.1016/j.biortech.2012.08.019> [accessed: 21 July 2014].

Rodriguez, L. and Preston, T.R. (1998) *Biodigester Installation Manual*, Ho Chi Minh City: FAO. <www.fao.org/ag/aga/agap/frg/Recycle/biodig/manual.htm> [accessed: 21 July 2014].

Roos, K.F., Martin, J.B.J. and Moser, M.A. (2004) *AgSTAR Handbook: A Manual For Developing Biogas Systems at Commercial Farms*, USA: Environmental Protection Agency <http://www.epa.gov/agstar/documents/AgSTAR-handbook.pdf> [accessed: 21 July 2014].

Simgas (2014) 'Press release Gesi550', [website – press release] Simgas <www.simgas.com/newsandoffers/news/press-release-gesi550/item74> [accessed: 21 July 2014].

Thiengburanathum, P. (2010) 'Thailand Swine Farm Biogas Implementation Programs/Policies', [Presentation] *Methane to Markets Partnership Expo*, New Delhi, India <www.globalmethane.org/expo-docs/india10/postexpo/ag_thiengburanathum.pdf> [accessed: 21 July 2014].

Viquez Arias, J. (2009) 'Case Study: Technical and economic feasibility of electricity generation with biogas in Costa Rica', *Dos Pinos Dairy Farmers Cooperative*, Alajuela, Costa Rica <www.adelaide.edu.au/biogas/anaerobic_digestion/casestudy.pdf> [accessed: 21 July 2014].

Vögeli, Y., Lohri, C.R., Kassenga, G., Baier, U. and Zurbrügg, C. (2009) *Technical and biological Performance of the ARTI compact Biogas plant for kitchen waste - Case study from Tanzania*, Dübendorf, Switzerland: Swiss Federal Institute of Aquatic Science and Technology (Eawag). < www.eawag.ch/forschung/sandec/publikationen/swm/dl/Voegeli_2009.pdf> [accessed: 21 July 2014].

CHAPTER 8
Ancillary equipment

Abstract

Ancillary equipment used with a biogas plant includes slurry and gas-handling equipment. Inlet slurry of dung and water can be mixed using various types of machine. Food waste often requires a pre-digester before it is added to the slurry. Outlet slurry should be used as compost. Gas-handling equipment includes pipes, valves, manometers, stoves, and lights. Gas pipes can be made of materials such as metal (steel or copper) and plastic (HDPE, PVC, etc.). The best valves to use for gas are ball valves. Gas stoves and lights need to be carefully designed and tested. Biogas can also be used in engines with suitable ancillary equipment.

Keywords: biogas, anaerobic digestion, ancillary equipment

Biogas equipment

Although the plant is the major part of a biogas system, it cannot exist by itself. A biogas plant must be fed, usually daily, with suitable feed; spent slurry must be removed and the biogas taken to where it is required. Suitable equipment must be provided to allow customers to use the gas, either in domestic appliances, such as stoves and lights, or in engines that drive useful machinery.

Commercial companies in India were keen to become involved in making biogas equipment, such as burners and lights, for the Khadi and Village Industries Commission (KVIC) programme. Such equipment has been available for the market in India ever since. The quality of this equipment is variable because it was designed on an *ad-hoc* basis. More recently academic institutions have been involved in improving equipment designs. The Nepal programme began by providing equipment imported from India. However, some of the required equipment was fairly quickly developed and made locally at a much lower cost. Locally made equipment can be serviced more easily and improvements made as required. However, the locally made equipment has been of a lower standard. In Nepal and India, a biogas plant is usually supplied as a package, which includes the appropriate ancillary equipment to fulfil the task for which it was purchased.

When Chinese peasants were responsible for setting up their own plants, they usually had to make their own stoves and lights. These were often of poor quality and low efficiency. Since the Ministry of Agriculture took responsibility for biogas standards, commercial companies in China have been making biogas equipment that is of a much higher standard.

Slurry-inlet handling

The bacteria in a digester cannot easily reach foodstuffs at the centre of lumps of animal dung or vegetable matter. For efficient digestion, the feedstock must be mixed with water into a smooth slurry before it is put into the digester. A masonry pit is usually provided at the entrance of the slurry inlet-pipe, where this mixing can be done. A plug, made of wood or concrete, is used to close the pipe until the feed is ready.

In Nepal and India cattle dung is usually mixed with water by hand. This is a culturally acceptable practice, but it is not recommended on health grounds because animal dung often contains intestinal parasites, such as amoeba and giadia. A simple tool was developed by Development and Consulting Services

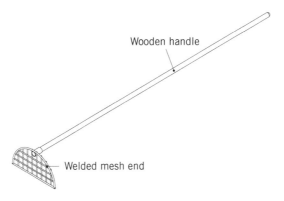

Wooden handle

Welded mesh end

Figure 8.1 Hand tool used to mix cattle dung slurry

(DCS) that looks like a hoe (Fulford, 1988), with the blade made of wire mesh, similar to a large potato masher (Figure 8.1). This can be used to mix the slurry and to break up lumps. The tool can also be used to remove strands of straw and woody material.

Larger biogas plants in Nepal and India used a mixing machine, designed by Khadi and Village Industries Commission (KVIC), based on a machine used to break up pulp for paper making (Bulmer *et al.*, 1985). It consisted of an elliptical masonry trough with a horizontal paddle-wheel beater set in it (Figure 8.2). A coarse sieve removed straw and other fibrous material. The machine worked fairly well and was popular, despite its tendency to splash slurry over the operator.

More recently, a mixing machine was developed by the Gobar Gas Company (GGC) (Devkota, 2001) in Nepal that uses a vertical shaft, with two mixer paddles at the bottom, within a cylindrical masonry pit (Figure 8.3).

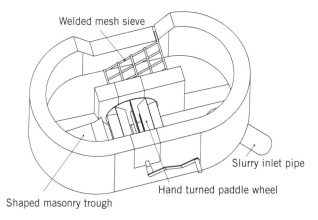

Welded mesh sieve

Slurry inlet pipe

Hand turned paddle wheel

Shaped masonry trough

Figure 8.2 KVIC mixing machine used to mix cattle dung slurry

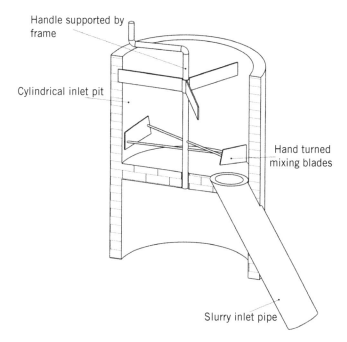

Handle supported by frame

Cylindrical inlet pit

Hand turned mixing blades

Slurry inlet pipe

Figure 8.3 GGC mixing machine used to mix cattle dung slurry

The paddles are turned by hand using a handle fitted to the top of the shaft. The machine acts in the same way as a food mixer and breaks up lumps.

A problem with using steel tools to mix dung is that the volatile fatty acids can cause corrosion of the metal. The metal parts must be painted with a good-quality paint. Such tools may need to be repainted each year and may need to be replaced after they are used for a few years.

The amount of water that is mixed with dung can vary, depending on the moisture content of the dung. Generally water is mixed with the dung to allow it to flow easily down the inlet pipe. In many cases a mixture of one part dung to one part water is used because it is easy to measure. If the inlet and outlet pipes are made large enough, it is possible to use undiluted dung in a biogas plant (Shyam, 2001).

Any material that enters a biogas plant from a latrine is usually piped directly into the digester, without mixing. When pig dung is used as a feedstock, as often happens in China, it is also fed in unmixed. These feed materials are often much less solid than cattle dung, so are broken down more easily.

The dung from other animals may need mixing, depending on its consistency. Dung from animals, such as horses, camels, and elephants, that eat grass and other vegetable matter, will need to be thoroughly mixed. Long fibrous strands are often present in the dung from these animals that needs to be removed from the slurry. These strands can cause blockages in the pipes and also float to the surface of the slurry to form a scum layer. Dung from goats and sheep can produce good biogas (Kanwar and Kalia, 1993), but the pellets need to be chopped or broken up and thoroughly mixed with water to form a slurry.

When food waste is added to a single-stage digester, it should be chopped up to form a slurry. Domestic food-waste biogas plants require about 1 kg of dry matter a day, which means 2 kg to 3 kg of wet material. This can be ground up in an old food blender or in a hand-powered mincing machine. VK NAREDEP in India has tested the use of a food waste-disposal unit, which is designed to grind up food waste so it can be washed down the drain. This does use electricity, but grinds the food waste very finely.

Food waste can be left to aerobically break down for a day or two before it is fed to the digester. For a domestic biogas plant, this does mean leaving the waste in a bucket with a lid, which can generate odours of rotting food, so it needs to be kept in a place where it will not attract flies or other pests and not be offensive to people.

Larger food-waste digesters can use a pre-digester. Designs used in India range from a 'soaking pit', a larger version of the aerobic bucket, to a specially designed solar-heated pre-digestion system (Shah, 2006). Faster pre-digestion occurs if liquor from the effluent of the main digester is recycled and mixed with the feed material. This first hydrolysis stage also works faster at 55 °C, so some form of heating (solar or otherwise) enables this to occur. Food waste can be chopped before it is added to a pre-digester.

A pre-digester can be designed so the food waste is flooded with recycled liquor, or as a leach bed, where the liquor trickles through the food waste. Care must be taken with a leach bed predigester to ensure that there are no lumps of material that the liquor cannot flow through.

Slurry-outlet handling

The slurry from a biogas digester is seen as valuable compost. It does need to be stored until it can be used. As described in Chapter 5, it can be stored in a slurry pit, as liquid; it can be left to dry; it can be added to dry biomass material and composted; or it can be composted and the composted material can then be vermi-composted.

Space close to the location of the digester needs to be allocated for a slurry pit and for a compost heap and vermi-compost unit, if they are to be used. If there is a lack of space close to the outlet of the digester, the slurry can be collected in containers and carried to a slurry pit or compost heap.

Slurry that is stored as a liquid needs to be put in a container, similar to a watering can, so it can be spread on the fields as a liquid fertilizer (Le, 2008). Some farmers in Vietnam mix the slurry in a pond with irrigation water and pump it to their crop fields.

Dry composted slurry can be carried in baskets or bags and spread over the fields in the same way as it done for inorganic fertilizer. It can be scattered over the ground by hand or with a shovel, to ensure it is evenly spread. Larger amounts can be spread mechanically.

Gas piping

Gas needs to be transported from the gas-storage container to where it is to be used and pipes usually convey the gas. Pipes are often laid underground to keep them out of the way. Biogas contains moisture, which can condense in the pipeline and cause blockages. Biogas pipelines should contain drainage points, towards which the pipe should slope, so that this water can be removed. More details of the use of gas pipes can be found in Appendix 4.

Gas pipes for small-scale biogas plants come in a range of fixed sizes. Older systems were measured in imperial units and were of ½ inch, ¾ inch and 1 inch nominal diameter. In many places, these have been replaced by metric sizes: 15 mm, 20 mm and 25 mm. These diameters are defined by international standards, but the actual sizes of pipes available in different markets may vary. The thickness of the walls of the pipe also varies, depending on the pressure for which the pipe is designed. The gas pressure from small biogas plants is low compared with the pressures for which pipes are usually specified, so standard pipe sizes can usually be used.

Gas pipes can be made from a range of materials, usually either metal or plastic. Older systems used galvanized iron pipes (steel pipes coated with zinc) although this choice is now seen as expensive. Galvanized pipes usually come in 2 m or 3 m lengths, with threads at the end. These can be screwed into sockets to join the pipes together. Other fittings, such as 'T' pieces and elbows are also available, so the pipe runs can be designed to suit the layout required. Specialized tools are available to bend and thread the pipes. Where the pipes are screwed into pipe fitting, such as sockets, the threads must be carefully sealed against gas leaks, using suitable jointing methods (see Appendix 4).

A growing range of different types of plastic pipe is now available. Good quality plastics are made from high-density polyethylene (HDPE) or PVC, although a lower-quality plastic polypropylene random copolymer (PPR) is also used. Low-quality pipes will quickly degrade and crack and leak. Plastic pipe comes in long rolls and can be easily cut to length. HDPE and PPR pipe can either be butt-welded or used with internal or external fittings. PVC pipe can be connected with pipe fittings that are solvent welded to the pipe lengths.

Plastic pipe is more easily damaged than metal pipe, so it needs protection. If used underground, it can be placed in a trench with a layer of sand around it (about 300 mm) to reduce the risk of damage by stones. This will also reduce damage by rodents (rats and porcupines), which often gnaw plastic pipes. If plastic pipe is used above ground, it should be protected from the sun.

The pipe diameter required depends on the length of pipe run required, the flow of gas and the maximum pressure drop that can be tolerated (see Appendix 4). The pressure from a floating drum is low (400 Pa to 800 Pa, which is about 40 mm to 80 mm water gauge), so the permissible pressure drop is also low (100 Pa or 10 mm WG). This means that small pipe diameters can only be used for short pipe runs (e.g. a 16 mm pipe used on a small domestic plant is limited to about 70 m, see Appendix 4). The pressure from a dome plant is higher (1,000 Pa to 12,000 Pa, 100 mm WG to 1,200 mm WG), so the pressure drop along the pipe can be much higher (500 Pa or 50 mm WG). This means longer lengths of smaller pipe can be used (up 350 m for the above example).

If a plastic gas pipe is used above ground, it can be supported so that the first section slopes back to the biogas plant. Any water that condenses in the line will then run back into the digester pit. There is a danger that above-ground gas pipes will form a series of loops between successive supports. As far as possible this should be avoided to prevent water condensing in the lower part of the loops and blocking the pipes. Many plastic pipes, if used above ground, will degrade in the ultraviolet from sunlight, so may need to be replaced after a few months. Pipes made from PVC and soft polyethylene are most likely to be affected by sunlight. HDPE pipes appear to last longer under such conditions.

If the gas line is put underground, it should be laid so that it slopes towards one or more low points at which water-draining points are located. The slope should be steeper than 1:100. There are different ways to remove the water, such as manual drain valves or self-draining devices (see Appendix 4).

Gas valves

Valves are used to turn the gas supply on and off. A main gas valve is required close to the plant, so that the gas line can be turned off for maintenance. Each appliance needs its own gas valve. Water taps are not suitable. Globe valves, which use a rubber washer, require fluid pressure to lift the washer from its seat. Gate valves use a tapered metal disc (the gate) that is pushed into a tapered slot by a rotating spindle, but they are not reliable. Users can over-tighten and break them and the gate wears out and leaks.

The cheapest gas valves are plug cocks, or quarter-turn gas valves. They use a tapered plug with a hole in it that fits in a tapered hole in the body (Bulmer *et al.*, 1985). The hole can be turned so it aligns with the gas pipe connected to the valve to allow gas to flow, or it can be turned so the hole is at right angles to the pipe (Figure 8.4). The tapered plug is held tight in the tapered hole with a spring. The plug tends to wear and become loose over time, allowing the gas to leak, so this type of valve needs to be replaced after a few years of use.

The best design of gas tap is a ball valve (Fulford, 1988). This uses a ball with a hole in it, which can be aligned with the pipe or rotated through a right angle to turn off the flow (Figure 8.5). The ball fits into spherical seals that press against the ball so it does not leak. Ball valves are mainly used as shut-off valves and are less effective in controlling the flow of the gas to an appliance.

Another type of valve, which is used in devices such as gas lights, is the needle valve (see Figure 8.6). This valve uses a tapered needle which can be

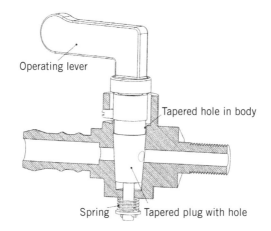

Figure 8.4 Tapered gas tap (or plug cock) – cross-section view to show working

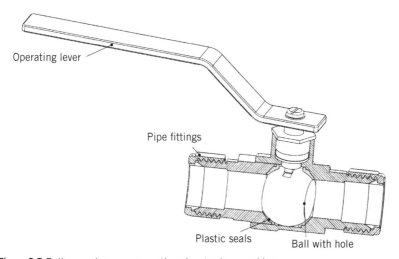

Figure 8.5 Ball gas valve – cross-section view to show working

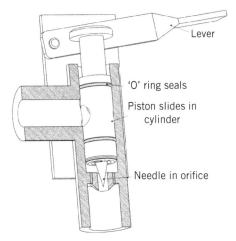

Figure 8.6 Needle gas valve – cross-section view to show working

moved in and out of a round hole to control the flow of gas. The advantage of a needle valve is that the gas flow can be increased steadily, while plug cocks and ball valves tend to be either on or off.

Valves need to be of materials that are resistant to corrosion. Since valves have moving parts, galvanized iron is not a suitable material because the zinc coating would rub off. Stainless steel is the best material to use and is the material from which the balls in a ball valve are usually made. Brass is also a suitable material and is often used to make the body of the valve. Biogas valves made from plastic are now available from China, but these will be less robust than valves made of metal that have plastics just used for the seals around the ball.

Manometers

A manometer is a simple pressure gauge. A manometer can be made from a transparent tube made into a 'U' shape, with a loop at the bottom. The whole pipe can be flexible, in which case the two sides of the U need to be supported so they are as straight as possible. Alternatively, the two straight sections can be rigid linked with a flexible pipe between them. The U tube is half-filled with water, so the pressure of the gas on one side, pushes the water up the other side. The two straight sides need to be long enough (up to 1.2 m for fixed-dome plants), so the water is not pushed out of the U tube. The water can be coloured with a water-soluble dye or ink, so the levels can be seen more easily.

The main use of a manometer is to check the gas pipeline for leaks (see Appendix 4). It is also used with fixed-dome plants to give the user a measure of how much gas is left in the plant. As gas is used from the dome, the gas pressure reduces.

Gas burners

Biogas has different combustion properties to pure methane (natural gas), so requires specialist burners for its use (Khandelwal and Gupta, 2009). The overall design of a gas burner is similar, whether it is used on natural gas or biogas, but details of the way the gas is mixed with air and the size and shape of the burner ports will be different. It is possible to adapt a natural gas stove to work on biogas, but the work needs to be done by someone who understands the best way to do it. LPG stoves are very different, as the pressure from an LPG bottle with a pressure-reducing valve is much higher. Adapting an LPG stove to run on biogas is difficult and not usually recommended.

Biogas stoves have been manufactured in India since the KVIC programme was established in the 1970s. The original biogas programme in China encouraged the use of locally made cheap stoves (van Buren, 1979), but the quality was very poor. Biogas stoves are now available from Chinese manufacturers, based on standards issued by the Chinese government in the 1980s (Chen *et al.*, 2012). A biogas stove was designed for the Nepal biogas programme by Development and Consulting Services DCS (Fulford, 1988) and has been manufactured by local Nepali companies for the biogas programme (BSP Nepal). Biogas programmes in countries such as Cambodia, where the SNV Netherlands Development Organisation was an advisor, adapted the DCS design of stove.

The DCS stove (Figure 8.7) illustrates the different components of a stove. Biogas is fed through a jet, which controls the flow. Most stoves are designed for a gas pressure of 750 Pa, which is the pressure from a floating drum plant. A typical domestic stove uses about 0.45 m^3 of biogas per hour, giving a heat output of 10 MJ per hour (2.8 kW), which is sufficient to bring 1 litre of water to boiling in four minutes (Appendix 5 gives details of stove design).

Gas enters the stove through a jet and is mixed with 'primary' air in a mixing tube. The mixture is fed to flame ports, where it burns with 'secondary air'. The supply of primary air controls the size of the flame; it is often controlled by varying the area of the air-supply ports. The burner ports are often arranged in a circle to suit normal cooking pots. The secondary air must flow freely around the burner ports and the combustion products be allowed to escape.

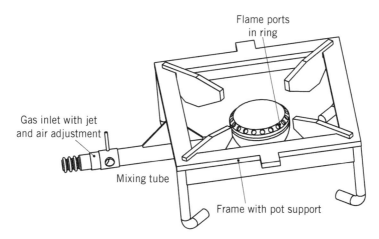

Figure 8.7 DCS biogas stove

The distance between the flames and the pots is important to achieve good heat transfer, so the frame of the stove must be designed to support the pots in the correct position. Stoves are often connected to the gas supply with a flexible pipe to allow them to be moved. A gas tap can be mounted on the kitchen wall, at the end of the fixed gas pipe, to allow the gas supply to be shut off if the stove is disconnected for cleaning. A gas tap can also be included as part of the stove to allow the flow of gas to be altered, allowing the heat out put of the stove to be adjusted.

Tests on the DCS stove, and other stoves made to a similar pattern, showed low efficiencies (CES, 2004, 2001). Stove design theory (Fulford, 1988) suggests that the total flame-port area is too small. Commercial biogas stove designs often use a toroidal (or donut) shaped burner manifold, so that flame ports can be placed on the inside of the toroid, as well as the outside (Figure 8.8).

Stoves made by commercial manufacturers are available from India and China, based on their standard natural gas and LPG designs. These stoves usually have a metal cover around the frame, which makes them easier to clean. These stoves are more efficient and provide better heat than those made by many biogas programmes (Khandelwal and Gupta, 2009).

Gas lights

Biogas lights are inefficient, expensive and need regular servicing. However, where electricity is not available, there is a good demand for such lights from owners of biogas plants. The principle of a biogas light is similar to that of a kerosene pressure lantern and the light is as good. A typical gaslight uses 0.09 m^3 to 0.18 m^3 per hour and generates similar light to a 40 W or 60 W incandescent electric bulb.

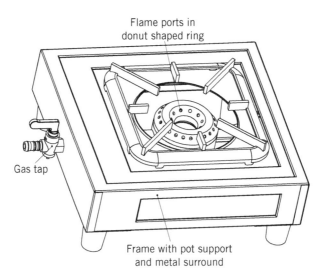

Figure 8.8 Typical Indian commercial biogas stove

The light burns biogas to heat an incandescent mantle that glows brightly when heated. The burner uses a jet to control the gas flow, a mixing tube that mixes the gas with air, and a burner port arrangement (the nozzle) that directs the flames onto the mantle. The burner is usually fitted with a regulator (a needle valve) to control the jet size. The air-inlet port can also be fitted with a restrictor to control the airflow. Careful adjustment is often needed to get the flame to a high temperature to get the best light.

A light was designed and made by DCS (Bulmer *et al.*, 1985). It did not work as well as the commercial lamps, but the drawing shows the different parts of the light (Figure 8.9). The nozzle is usually made from ceramic to withstand the high temperature in the flame. The primary air supply is close to 100 per cent to give a flame that burns close to the nozzle. The nozzle often has a central flame port surrounded by a ring of smaller ports, to give good flame stability. The mantle is made of a fine silk or rayon mesh soaked with a mixture of rare earth oxides, thorium and cerium. The mantle is tied around the end of the nozzle and spread into a dome shape, so the flame can heat it to a high temperature. When the lamp is first lit, the fabric burns away leaving a fragile mesh of rare earth salts that glow brightly when they are heated to a sufficient temperature.

Gas lamps are usually mounted vertically, often with the mantle facing downwards, and have a reflector to direct the light downwards. The reflector also directs the heat from the burner downwards. The mantle is covered with a glass globe to stop the flame being affected by draughts and to protect the fragile mantle from damage by insects.

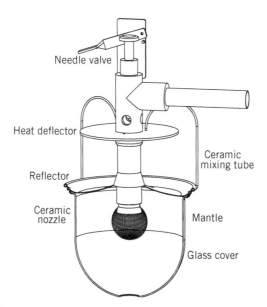

Figure 8.9 DCS light – cross-section view to show working

Various biogas programmes have attempted to make their own biogas lights (Khandelwal and Gupta, 2009), but the quality was poor. There are commercial companies that have been making biogas lamps in India to meet the demand for lamps encouraged by the KVIC biogas programme, but they always struggled to balance quality and price. There has been a similar relationship between the demand for high-quality products from the Chinese biogas programme and local commercial companies.

Use of biogas in engines

The gas from larger biogas plants can be used to run internal combustion engines (Mitzlaff, 1988). Although the heating value of biogas (17.8 MJ kg^{-1}) is much lower than that for gasoline (44.9 MJ kg^{-1}), an engine run on biogas is only derated by about 6 per cent (reduced energy output from the engine). The overall efficiency is about the same. However, biogas is much less dense than gasoline, so an engine needs to have a higher compression ratio (14:1 up to 21:1) in order to get sufficient gas into the engine. Normal gasoline (or Otto cycle) engines are limited to a compression ratio of 10:1. Higher pressures can cause pre-ignition or knocking when used with gasoline.

There are two ways to run an engine on biogas: as 'duel-fuel', which uses compression ignition (Makareviciene *et al.*, 2013), or on biogas alone, which requires a spark for ignition (Porpatham *et al.*, 2012). A duel-fuel engine draws in biogas into the cylinder along with air and then injects a proportion of diesel, or biodiesel, during compression, to ignite the mixture. The proportion depends on the size of engine: a small engine (5 kW) requires about 20 per cent diesel or biodiesel, while a very large engine (5 MW) requires only about 5 per cent. Spark-ignition engines run on methane or biogas are usually adapted diesel engines, with the fuel injector replaced with a spark plug. These are commercially available as gas engines for use with natural gas. However, gas engines that are run on biogas need to be more rugged, so not all gas engines are suitable for use with biogas.

Engines smaller than about 5 kW are available, but tend to be much less efficient, so are seldom used with biogas. A 5 kW engine requires 2.8 m^3 of biogas per hour when run in dual-fuel mode and 3.5 m^3 of biogas per hour when run in spark-ignition mode (Fulford, 1988). This means that a larger size of biogas plant is required than is used for domestic purposes and a larger amount of feed material is also required. For example, if cattle supply the dung for the feed material, at least 50 animals are required.

Large engines can be used to drive electricity generators or to provide mechanical power. They can be used to drive water pumps for irrigation or food processing machinery, such as flourmills and rice hullers. Various devices are used to mix the gas and air in both dual-fuel and spark ignition engines.

If biogas is used in a dual-fuel slow running engine, it can be mixed with the air supply by making a simple mixing chamber between the air filter and the air-inlet pipe. The gas supply is fed through a pipe into the mixing chamber and the flow controlled by a gas valve. Most diesel engines use a mechanical or electronic governor that controls the diesel flow depending on the speed of the engine. As the gas valve is gently opened and the engine speeds up in response to the extra energy in the cylinders, so the governor reduces the diesel flow in response. When the gas flow is further increased, the amount of diesel is reduced to the point where it is inadequate to give good combustion. The engine will begin to run badly, so the gas supply is reduced until the engine runs smoothly again. This is the optimum setting for running the engine in dual-fuel mode.

Faster running dual-fuel engines and spark-ignition engines require a more accurate mixture of biogas and air, so a gas carburettor is required. The air pressure is reduced as it passes through a venturi to draw gas into the air stream and mix with it (see Figure 8.10). The gas flow is determined by the pressure of the gas supply, which must be controlled. The details of the design of the gas carburettor will depend on the particular engine with which it is used.

It is often recommended that hydrogen sulfide is removed from the biogas before it is burnt in engines. The sulfur burns in air to generate sulfur oxides which form corrosive acids when mixed with the water, which is also produced by combustion, when it condenses (Caterpillar, 2013). This acid can corrode parts of the engine. There are both biological and chemical systems that can be used to remove hydrogen sulfide (Wellinger and Murphy, 2013). There are many microbes that process hydrogen sulfide and convert it to sulfur or sulfates (Syed *et al.*, 2006). Some use light for the process and other require a small amount of air to be bled into the filter. The most common chemical method uses a reaction with iron to form iron sulfide (Abatzoglou and Boivin, 2009). A filter can be filled with wet, rusty steel shavings with which the hydrogen sulfide reacts. The steel needs to be replaced periodically, as the rust is replaced with an insoluble layer of black iron sulfide.

Engines can be used with biogas that has not been cleaned of hydrogen sulphide, but the corrosive effects of the combustion products will reduce the lifetime of the engines.

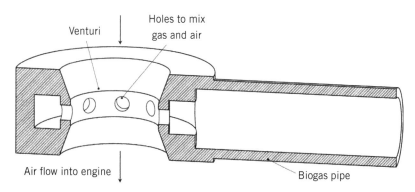

Figure 8.10 Gas venturi – cross section view to show function

Use of ancillary equipment

The behaviour of biogas equipment will be influenced by the type of biogas plant that is used to generate the gas.

The gas pressure from flexible-bag and tent plants is very low (a few Pa). This means that the danger of leaks from the gas pipe and valves is low. However, the flow of gas through burners is also low, so they will not generate the heat required. The flow of biogas along pipes will also be much reduced; the pressure drop depends on the pipe size and the gas flow rate. Larger diameter pipes can overcome this problem. Weights can be placed on some designs of gas bag to increase the pressure, but this is not possible if increased pressure causes effluent slurry to be forced out of the bag. Some designs of bag plant use a separate gas-storage bag. When the biogas is used, a valve in the pipe connecting the biogas plant to the storage bag is closed. The pressure in the gas-storage bag can then be increased, by placing weights on it or by wrapping elastic bands around it.

The gas pressure from floating drum plants is roughly constant. Biogas burners are designed to work at around 700 Pa, so they should work well when run with this type of plant. The gas flow can be adjusted by using the control valve on the gas burner. If reduced power is required, for example to allow food in a cooking pan to simmer once it has been brought to boiling, the control valve can be closed by a suitable amount.

The gas pressure from fixed-dome plants is much higher when the plant is full of gas (up to 12,000 Pa), so the gas control valve will need to be almost closed to run a burner correctly. When gas is used, the pressure in the plant will drop, so the gas valve may need to be adjusted to keep a constant power output. The high gas pressure will mean that gas pipes are more likely to leak, so extra care must be taken to ensure joints are well sealed.

Engines require even lower gas pressure, so a gas-pressure control valve is needed for most types of biogas plant. Commercial pressure regulators are available, which use a rubber diaphragm linked to a needle valve. Such regulators are used with LPG and are designed to reduce gas pressure from over 1,200 kPa to about 2.75 kPa. The outlet pressure is set with a spring, so it can be adjusted with a screw that alters the spring tension. Commercial regulators to reduce pressure from 12,000 Pa to less than 700 Pa are more difficult to find. A low-cost regulator that can be made from two buckets and standard pipe fitting is shown in Figure 8.11. The lower bucket is filled with water, so that the gas enters the space between the buckets. The control pressure is adjusted by adding weights to the top of the upper bucket.

The easiest way to obtain biogas equipment such as valves, burners and lights is to purchase them from India or China. Many equipment suppliers belong to international trade associations, which allows orders to be placed online.

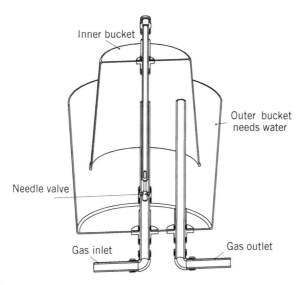

Figure 8.11 Simple low-cost gas-pressure regulating valve

References

Abatzoglou, N. and Boivin, S. (2009) 'A review of biogas purification processes', *Biofuels, Bioprod. Bioref.* 3: 42–71 <http://dx.doi.org/10.1002/bbb.117> [accessed 21 July 2014].

Bulmer, A., Finlay, J., Fulford, D.J. and Lau-Wong, M.M. (1985) *Biogas: Challenges and experience from Nepal Vols. I and II*, Butwal, Nepal: United Mission to Nepal. <www.kingdombio.com/Biogas-vol-I.pdf> [accessed 21 July 2014].

van Buren, A. (1979) *A Chinese Biogas Manual*, London: Practical Action Publications (IntermediateTechonology Publicatons). <HYPERLINK "http://developmentbookshop.com/chinese-biogas-" http://developmentbookshop.com/chinese-biogas- manual-pb> [accessed 22 July 2014].

Caterpillar (2013) *Maximizing Productivity of Biogas Engines*, AgStar Environmental Protection Agency, Washington DC. <www.epa.gov/agstar/documents/conf13/Maximizing%20Productivity%20of%20Biogas%20Eng ines,%20Mike%20Devine.pdf> [accessed 21 July 2014].

CES (2001) *Efficiency Measurement of Biogas, Kerosene and LPG Stoves*, Pulchowk, Lalitpur, Nepal: Institute of Engineering, Tribhuvan University. <www.snvworld.org/en/download/publications/efficiency_measurement_of_biogas_kerosene_an d_lpg_stoves_nepal_2001.pdf> [accessed 21 July 2014].

CES (2004) *Efficiency Measurement of Biogas Stoves*, Pulchowk, Lalitpur, Nepal: Institute of Engineering, Tribhuvan University. <www.snvworld.org/en/download/publications/efficiency_measurement_biogas_stoves_nepal_2004.pdf> [accessed 21 July 2014].

Chen, L., Zhao, L., Ren, C. and Wang, F. (2012) 'The progress and prospects of rural biogas production in China', *Renewable Energy in China* 51: 58–63 <http://dx.doi.org/10.1016/j.enpol.2012.05.052> [accessed 21 July 2014].

Devkota, G.P. (2001) *Biogas Technology in Nepal: A sustainable source of energy for rural people*, Kathmandu: Mrs. Bindu Devkota, Maipee.

Fulford, D. (1988) *Running a Biogas Programme: A handbook*, London: Practical Action Publications (Intermediate Technology Publications). <http://developmentbookshop.com/running-a-biogas-programme-pb> [accessed 22 July 2014].

Kanwar, S.S. and Kalia, A.K. (1993) 'Anaerobic fermentation of sheep droppings for biogas production', *World Journal of Microbiology and Biotechnology* 9: 174–175.

Khandelwal, K.C. and Gupta, V.K. (2009) *Popular Summary of the Test Reports on Biogas Stoves and Lamps Prepared by Testing Institutes in China, India and the Netherlands*, SNV Netherlands Development Organisation.The Hague, Netherlands <www.snvworld.org/en/Documents/Biogas_stoves_and_lamps_test_report_2009.pdf> [accessed 21 July 2014].

Le, T.X.T. (2008) *Bio-slurry Utilization in Vietnam, Hanoi, Vietnam: The Biogas Program for the Animal Husbandry Sector of Vietnam*, MARD Vietnam. <http://biogas.org.vn/vietnam/getattachment/Thu-vien-phim,-anh/An-pham/Bao-cao-tinh-hinh-su-dung-phu-pham-khi-sinh-hoc-ta/Bao-cao-tinh-hinh-su-dung-phu-pham-khi-sinh-hoc-tai-Viet-Nam.pdf.aspx> [accessed 21 July 2014].

Makareviciene, V., Sendzikiene, E., Pukalskas, S., Rimkus, A. and Vegneris, R. (2013) 'Performance and emission characteristics of biogas used in diesel engine operation', *Energy Conversion and Management* 75: 224–233 <http://dx.doi.org/10.1016/j.enconman.2013.06.012> [accessed 21 July 2014].

Mitzlaff, K. von (1988) *Engines for Biogas*, Eschborn: German Appropriate Technology Exchange. <www.gate-international.org/documents/publications/webdocs/pdfs/g36ene.pdf> [accessed 22 July 2014].

Porpatham, E., Ramesh, A. and Nagalingam, B. (2012) 'Effect of compression ratio on the performance and combustion of a biogas fuelled spark ignition engine', *Fuel* 95: 247–256 <http://dx.doi.org/10.1016/j.fuel.2011.10.059> [accessed 21 July 2014].

Shah, O.P. (2006) *Sustainable Waste Processing in Mumbai: Using the Nisargruna technology*. MSc Thesis. Borås, Sweden: Hogskolan i Borås. <www.kingdombio.co.uk/ShahThesis.pdf> [accessed 21 July 2014].

Shyam, M. (2001) 'A biogas plant for the digestion of fresh undiluted cattle dung', *Boiling Point* 47: 33–35 <www.hedon.info/View+Article&itemId=9164> [accessed 22 July 2014].

Syed, M., Soreanu, G., Falletta, P. and Béland, M. (2006) 'Removal of hydrogen sulfide from gas streams using biological processes - A review', *Canadian Biosystems Engineering* 48: 2.1-2.14 < http://engrwww.usask.ca/oldsite/societies/csae/protectedpapers/c0436.pdf > [accessed 22 July 2014].

Wellinger, A. and Murphy, J. (2013) *Biogas handbook: Science, production and application*, D Woodhead, Cambridge, UK Publishing Series in Energy <www.iea-biogas.net/biogas-handbook.html> [accessed 21 July 2014].

CHAPTER 9
Using biogas plants

Abstract

The process of building a biogas plant involves a series of decisions: the design of plant to be used; the feedstock available and therefore the size of the plant; the location of the plant and its distance from the feedstock supply; and where the gas is to be used. Starting a plant involves adding the right microbes and allowing the population to be built up. Running a plant is a matter of feeding it regularly and avoiding contaminants.

Keywords: biogas, anaerobic digestion, ancillary equipment

Choice of biogas plant design

There are several factors that must be considered when choosing a design of biogas plant, including cost, lifetime, and ease of construction. However, a major factor in determining the particular design to use is whether there are trained workers available who are able to build it. All plant designs require a degree of skill to make, especially the underground dome systems. Poor quality workmanship can easily result in failure of the digester.

Use of trained staff

The need for properly trained staff was illustrated in the experiences of using versions of the Chinese fixed-dome in India and Nepal using a domed roof made from concrete. This version, with a concrete dome, is called the *Janata* ('people's') plant. An Indian NGO, Action For Food Production (AFPRO), introduced the design in India and trained people to make it. The concrete roof was made over a mould, but the quality of the moulds used seems to have been variable. In a study of the performance of 615 biogas plants, 43 per cent of the Janata plants had failed (Pal *et al.*, 2002). A similar design is made in Nepal, again with a concrete roof dome, poured over a mud mould. But the construction workers in Nepal are properly trained to make this design, so the failure rate was less than 5 per cent (Ashden, 2005a).

In India, the Deenbandhu ('friend of the poor') plant with a dome made from brick, has proved much more reliable, with 75 per cent of the 476 plants of that type working well. There seems to have been more thorough training of construction staff by groups such as the Khadi and Village Industries

http://dx.doi.org/10.3362/9781780448497.009

Commission (KVIC) and Action For Food Production (AFPRO), and these skills are passed on to others in the construction groups. SKG Sangha is one NGO group which has demonstrated the reliability of this design (Ashden, 2007a). GTZ (a German Aid Agency, now called GIZ) has also run courses on how to build this type of plant in East Africa, so there seems to be a skill pool in that area of the world. Several groups have started building biogas plants in Kenya as a result of GTZ's courses in the past.

In Latin America, the main expertise seems to be in the manufacture of plastic bag or tent digesters. Groups in Costa Rica and Mexico have produced standard designs, which are being used in places such as Bolivia and elsewhere in the region.

Criteria for design choice

Where there is a choice of available designs, people tend to change the criteria on which they select a design once a programme becomes established. While low cost seems to be more important at the start of a programme, good quality and high reliability become more important over time.

In Vietnam, for example, several biogas plant designs have been offered. Initially, plastic-bag digesters were very popular; they were cheap and easy to install. However, users became disillusioned with the poor durability of these digesters and started looking for alternatives. The Vietnam Gardening Association developed the VACVINA (Hunt, 2009), which uses a rectangular concrete lined pit for the digester and a plastic bag for gas storage. The cost is higher, but the digester tank lasts much longer. Several thousand of these plants have been built. This plant uses a very low gas pressure, which is typical of plastic-bag digesters, but the project staff claim that it reduces the risk of gas leaks.

The Vietnamese Ministry of Agriculture and Rural development (MARD) has worked with Netherlands Development Organisation (SNV), to introduce a fixed-dome digester to Vietnam (Ashden, 2010a). They chose a design similar to the Deenbandhu, because bricks are easily available in Vietnam and staff had already been trained to build this design. Over 75,000 units had been made under this programme by the end of 2009. The Vietnamese experience suggests that people initially choose low-cost designs of biogas plant. However, as they recognize the benefits, many people will choose a more reliable and much longer lasting design, despite the higher costs.

Factors for design selection

If biogas plants are to be built in an area where there are no trained staff, the choice of plant design depends on a range of factors. Cost is often a major issue, because potential users want the cheapest solution, such as bag or tent digesters. However, the low reliability and short lifetime of these plants often lead people to look for better alternatives. The actual cost of different designs often depends on local factors, including the cost of labour and availability of suitable materials.

Of the different types of biogas plant considered in this book, the most expensive are those using a floating drum to collect the gas. Larger floating drums are usually made from steel, while small drums can be made from plastic. One major advantage of a floating drum plant is that the drum can be removed for cleaning the plant. This can be of benefit if the plant is fed with material containing lignin, such as food wastes. The food waste plants used in India tend to use floating drums. There are types of plant that are made completely of steel or of rigid plastic and these tend to be even more expensive. The advantage of small rigid-plastic biogas plants is that they can be easily moved when they are empty. The low cost of installation and the flexibility of where they can be installed are factors that offset the higher costs. For example, such plants can be placed on a flat roof or a large veranda (Ashden, 2006).

The fixed-dome designs are seen as cheaper than floating drum designs, because they are made mainly of masonry, which is cheaper than steel or rigid plastic. The typical cost of a fixed-dome plant is about 59 per cent of that of a floating drum plant of the same size (Khandelwal, 2008). The relative cost of different designs of fixed-dome plant depends on the cost of the materials used for construction in the locality where it is to be built. If bricks are easily available, then the Deenbandhu design is likely to be the cheaper option. If bricks are less easy to obtain, but aggregate for making concrete is easily available (for example in more mountainous areas), then a design using a concrete dome, such as the Gobar Gas Company (GGC) design, will cost less.

Flexible-plastic plants are seen as the cheapest to install, but the actual cost depends on the quality of the materials used to make the plant. The lifetime of a bag digester is related to the cost of the materials: low cost means short lifetime (Nazir, 1991), while a longer lifetime means higher costs.

Lifetime is another factor that determines the choice of design. Users come to rely on the gas that their plant produces, so a failure of the plant causes them problems. Bag plants are seen to have a short lifetime, because flexible plastics do degrade with time. Even high-quality plastic materials that are well protected will deteriorate with time. This is also true of rigid plastic materials used in floating drum plants, although the lifetime of these materials will be much longer than that of flexible plastic materials.

Steel gas drums will tend to rust, unless they are well protected with good quality paint. The process of anaerobic digestion generates volatile fatty acids (VFA) that can attack steel, especially in the presence of oxygen. The inside of the drum is free from oxygen, but the outside is immersed in the slurry and then exposed to air, as the gas drum fills with gas. Such drums need to be painted every five years, or more often, to avoid the steel rusting.

Fixed-dome biogas plants, if they are made by skilled people, have the longest lifetime. Detailed follow-up surveys of older plants have not been done, but most studies assume a lifetime of at least 15 years. A properly built fixed-dome biogas plant could last for 100 years; there are few parts in the plant that will degrade with time, apart from the gas pipework, which may need to be replaced more often. A rice godown (store) that was made as a brick dome in Patna in 1786, the Golghar, has lasted over 225 years and tourists are still able to climb to the top.

Stress analysis calculations were done on the GGC design of biogas plants used in Nepal. A 20 m³ plant was analyzed (Bulmer *et al.*, 1985) and the results suggested the dome had a safety margin of 10, i.e. it was 10 times stronger than it needed to be. A tractor could be placed above the backfill over the dome and the dome could easily take the strain. The experiences of the companies that have built 300,000 of these plants indicate that none of the domes have ever cracked, even those that are 30 years old. There were a few cases where the domes collapsed soon after manufacture, but these were caused by the use of cement that had spoiled because of damp.

The use of steel reinforcement is actually counter-productive in terms of the life of concrete. Steel bars can rust if moisture gets into the concrete over time. Rust causes the bars to swell, which can crack the concrete after a time. Using steel reinforcement will shorten the lifetime of a concrete dome. The lifetime will be determined by how fast moisture and air can penetrate into the concrete and cause the steel to rust and whether salts are dissolved in the moisture.

Fixed-dome plants with a concrete plug in the top of the dome can develop leaks around the plug, if it is not maintained carefully or the plug is removed frequently. The plug is usually sealed with clay, which needs to be kept moist to maintain a good seal. If the clay dries out, it can crack and allow gas to escape. Each time the plug is removed, there is a danger that the concrete of the plug or that of the edge of the hole can flake off. A thicker layer of clay can be used to fill the hole this leaves, but after a time, the hole and the plug may need to be remade in order to ensure that a good seal can be maintained.

Domestic plant sizes

Domestic biogas plant designs come in a wide range of sizes, depending on the amount of gas that needs to be generated and the amount and type of material available as a feedstock. Another major factor is the ambient temperature where the plant is installed; temperature affects the rate of gas production.

The way the working volume of a biogas plant is defined depends on the design. For a floating drum plant, the working volume is the volume of slurry in the digester pit. The slurry in a fixed-dome plant moves in and out from the digester pit into the reservoir. The working volume can be defined as the volume of the slurry that remains permanently in the digester pit, plus half that of the volume that flows in and out (i.e. the volume of the slurry reservoir). Both these measures are different from the total volume of the underground pit, which also includes a volume for permanent gas storage. A bag digester is usually partly filled with slurry; the upper portion is used for gas storage. It can be assumed that two-thirds of the total volume of the bag is filled with slurry (Davis and Preston, 1983), which defines the working volume of this digester. A tent digester uses the flexible membrane to contain the gas, so the working volume is the total volume of the trench used to contain the slurry. Some of the Indian programmes define the size of the plants by the amount of gas produced per day, so a plant with a working volume of 4.2 m^3 is called a 2 m^3 plant. However, some suppliers tend to exaggerate the gas production, so this is not a reliable measure of plant size.

Most domestic plants are used to generate cooking fuel. The smallest plant used for this purpose has a working volume of 1 m^3 and is usually fed with food waste. Food materials containing about 1.5 kg of dry matter in about 15 litres of water fed into a suitable plant each day can generate enough gas to cook food for a small family (Ashden, 2006). The equivalent plant using animal dung as a feedstock has a working volume of 4.4 m^3 and requires at least 40 kg of wet dung (6.4 kg dry matter) each day – animal dung generates less gas per unit weight of material. The smallest Deenbandhu digester has a working volume of 2.2 m^3 and can be fed with 20 kg wet dung a day (3.2 kg dry matter), but the gas generated is sufficient to cook meals for only one or two people each day (Satyamoorty, 1999).

The recommended plant sizes have reduced as the various biogas programmes have gained experience and realized that they needed to reduce the cost of plants. The original smallest KVIC floating drum plant had a working volume of 7 m^3. The early KVIC plants were built in northern India, while the more recent biogas programmes have worked in southern India, which has a higher average temperature. The smallest working volume of the fixed-dome plants built in Nepal by DCS was 7.9 m^3 (based on a total volume of 10 m^3), while at present the smallest plants built under the BSP Nepal programme have a working volume of 3.2 m^3 (based on a total volume of 4.2 m^3).

A digester with a working volume of 4.4 m^3 can be fed with the dung from 4 to 6 cattle or 8 to 12 pigs, to produce about 2 m^3 of biogas a day (Devkota, 2001; Satyamoorty, 1999). This amount of gas can run two biogas stoves for about two hours a day, which is sufficient to cook food for a small family. A larger family who also own a larger number of animals can use a larger size plant. Most programmes make a range of plant sizes, so that larger families can use more gas from more feed material.

Typical plant sizes from different programmes are shown in Tables 9.1 to 9.3. Different programmes seem to vary in the way they define their nominal plant sizes. The volume of permanent slurry is determined by calculating the volume of slurry in the plant when the gas storage volume is full. The working volume is defined as above (permanent slurry plus half of the slurry that moves in and out). The total volume of the digester pit will be larger, as it includes a volume for permanent gas storage, to prevent slurry entering the gas outlet pipe. The Nepal and Chinese programmes relate their nominal volume to the volume of the digester pit, while in South India, the nominal volume relates to the volume of gas expected to be produced each day, which is about half of the permanent slurry volume. Gas production depends on

Table 9.1 Plant volumes used in the BSP Nepal biogas programme [Devkota 2001]

Nominal volume (m³)	4	6	8	10	15	20
Permanent slurry (m³)	2.81	4.30	6.01	7.00	11.06	14.31
Working volume (m³)	3.23	4.84	6.73	7.93	12.37	16.31
Total volume of pit (m³)	4.25	6.06	8.22	10.12	15.08	21.62

Table 9.2 Plant volumes defined by the National Standard for biogas in China [NSPRC 1985]

Nominal volume (m³)	4	6	8	10
Permanent slurry (m³)	3.74	5.07	6.64	8.48
Working volume (m³)	4.05	5.48	7.19	9.18
Total volume of pit (m³)	4.52	6.13	8.05	10.30

Table 9.3 Plant volumes used for Deenbandhu plants in South India [Satyamoorty 1999]

Nominal volume (m³)	1	2	3	4
Permanent slurry (m³)	2.04	4.05	6.09	7.99
Working volume (m³)	2.23	4.40	6.62	8.69
Total volume of pit (m³)	3.24	5.78	8.47	11.08

temperature, but the ambient temperature in South India remains fairly high. The gas production figures that are used by the programme in China are much lower. The ambient temperature varies a great deal over the year and also in different places in China.

Larger plant sizes

Biogas plants of larger volume are used to provide enough fuel to provide gas for an institution (e.g. a school) or to run an engine. A plant with a working volume of 34 m³ can run a 3 kW (5 HP) engine for seven hours a day, using dung from 20 to 30 cattle (250 kg to 300 kg of wet dung) (Fulford, 1988). In India, most of the larger plants that have been built use a floating-steel drum. A typical working volume for such plants is 25 m³ (Heeb, 2009). The steel gas-drum for a plant with a volume of 6 m³ can be transported on the back of a truck. Individual plants up to 50 m³ working volume (Heeb, 2009) are being installed, but the larger steel gas-drums are more difficult to transport. They have to be welded together on site. If larger volumes of feedstock need to be processed, several 25 m³ or 50 m³ plants are used. The advantage of having several digesters is that one can be cleaned while the others are still being used. This type of biogas plant is used to process food waste and human sewage from schools, hostels and other institutions in South India to provide cooking fuel (Ashden, 2007b). Similar plants are used to process market wastes to provide gas to generate electricity, also in South India.

It is possible to build fixed-dome plants with a total volume as large as 100 m³ (Ashden, 2005b; Idan, 2008). Even larger plants are then formed by connecting together several of these 100 m³ or smaller plants. Such plants are used in Rwanda, Tanzania, and Kenya (Ashden, 2010b) to process sewage from schools and jails to provide fuel for cooking the inmates' food. A similar system is used in Ghana to process sewage from schools, colleges, hospitals, hotels, and other organizations to provide cooking fuel to be used by the organization (Idan, 2008). The sizes of these plants range from 60 m³ to 450 m³, although the larger plants are made as several smaller plants linked together. The largest such plant was built for a jail in Rwanda and has a total volume of 1,400 m³, and consists of 14 100 m³ plants linked together in two groups (Ashden, 2005b).

Biogas plants in climates with low ambient temperatures

The large biogas programmes (in China, India, Nepal, and other places in Asia) have proved effective where the local climate provides ambient temperatures that are normally above 28 °C. The designs that are available are not really suitable for being insulated and heated. In colder areas of countries that are running these programmes (such as at higher altitudes), people have devised ways in which the biogas plants can be run (see Chapter 6).

Site selection for biogas plants

Biogas salesmen and technicians need to be trained in the selection of the right site to use when they are installing a biogas plant. If an underground plant is to be built, the soil into which it is placed must be suitable. Soil that is very loose or sandy, or soil that is water-logged, is very difficult and expensive to dig. Also, if the ground is very hard and rocky, excavation can be time consuming and expensive. The use of a small plastic above-ground design may be more appropriate in these situations, if the supply of feedstock is suitable for such a plant.

A biogas plant needs to be close to the supply of feed material, such as the animal sheds, if animal dung is the main feedstock. The plant must also be close enough to where the gas is to be used; long gas pipelines are expensive. A biogas pit should be at least 15 m away from a drinking water source, such as a well, to prevent contamination (Fulford, 1988). There is a danger that tree roots can grow into a digester pit and crack the walls, so a potential site should be away from existing trees.

Many domestic biogas plant programmes encourage people to link a latrine to their plant, especially where there is no other provision for sanitation. The latrine is usually connected to a second straight feed pipe that goes directly into the slurry pit. The latrine floor must be at least 0.3 m above the slurry outlet (Fulford, 1988) to avoid flooding. The amount of water used to flush the latrine should be limited. These latrines do not use a 'U' bend between the pan and the digester pit, so require much less water than a western WC. Since the end of the pipe from the latrine ends well under the surface of the slurry in the pit there is minimal smell. Most latrines attached to biogas plants do not have a cistern. Water is carried to the latrine in a jug or other container; this enables the amount of water to be controlled.

In China, where the weather is often much colder than that in India, the biogas plant, pigsty and latrine are built very close together, with the digester often underneath the pigsty (the 3-in-1 system) (White, 2005). A variation is the 4-in-1 system (Yin, 2001), where the whole system is enclosed in a greenhouse (Cheng et al., 2011). The slurry outlet from the biogas plant is designed so that the slurry can be used directly on the crops in the greenhouse. Having the biogas plant underneath a greenhouse, with a pigsty in it, reduces heat loss and ensures the slurry remains at a high enough temperature to generate biogas. The pig dung can be easily collected and put in the mixing pit for the biogas plant, which can be sited in one corner of the pigsty.

Constructing a biogas plant

Once a design has been chosen for a biogas plant and a site has been chosen, the next step is constructing the plant. The work of building a biogas plant should be done by a trained team of installers who are part of a biogas extension programme. They have access to the right materials and equipment and are experienced in constructing the most appropriate design for an area.

Construction manuals are available for a range of different designs, such as Deenbandhu (Kudaravalli, 2014a; VK NARDEP, 1993), GGC (Devkota, 2003) and plastic-bag systems (Davis and Preston, 1983; Lüer, 2010). Individual groups can use these manuals to build their own designs, but they do not have the resources of good quality control and follow-up services that a biogas extension programme can offer.

Starting a biogas plant

A biogas plant requires a suitable population of microbes, which work together symbiotically to break down food materials to make methane and carbon dioxide. When a biogas plant is started, this population needs to be added to the digester and then allowed to grow and adapt. The best source of anaerobic microbes is from a working digester, although most of the microbes can be obtained from cattle dung. If a starter is used from a working digester, the proportion of starter to the main feed material determines how fast the digester can start. Only 5 per cent starter mixed with pig dung means a slow start, while 33 per cent starter in a digester will allow start up in only one third of the time (Maramba, 1978).

If cattle dung is the main source of feed material, the plant is first filled with a slurry of dung and water. The slurry should be well mixed, so that lumps are broken up. If the ambient temperature is above 28 °C, the microbes will naturally grow and start the digestion process. If the slurry is at a lower temperature, it should be warmed gently until it reaches the working temperature. Heating rates should be less than 5 °C per day, faster heating will cause the microbes to stop working. The optimum temperature for operation is 35 °C, and the microbes will grow and start working after a few days (up to a week). At lower temperatures, the process takes longer and more time needs to be allowed.

The first process that starts is hydrolysis, which consumes the oxygen in the air above the slurry. A manometer connected to the gas line will show a drop in pressure while this is happening. The first gas that is generated is carbon dioxide, which will not burn. This gas can be allowed to escape, as it flushes air out from the gas line. As the microbes become settled, the proportion of methane in the gas will increase, until it eventually can be ignited in gas appliances and will sustain a flame. Once this happens, the biogas plant can be fed with the planned feed material (animal dung or food residues). The amount of feed should be increased slowly up to the planned daily amount. Most biogas plants are fed once a day, so the initial feed rate should be about one quarter of the planned amount. This can be doubled to half after several days and then doubled again to the full amount after a few more days.

The main reason for a lack of gas being produced is the presence of gas leaks in the pipeline. The pipeline should have been carefully checked for leaks before any feed material is added (see Appendix 4). Once gas is being produced, it is worth checking the gas line again to ensure it has not been damaged. The main gas valve close to the plant can be closed and a manometer connected to the line checked to ensure the pipe holds pressure. Soapy water can be used to identify the position of any leaks.

The second main reason for a lack of gas is that the slurry is at too low a temperature. The ambient temperature can be monitored with a thermometer. If the plant is designed not to be heated, several weeks of patience may be required until the ambient temperature has naturally risen to above 28 °C. Once the microbial population has settled, it will operate at lower temperatures, although the daily gas production will be reduced.

Running a biogas plant

The microbial population in a biogas plant reacts poorly to changes, especially if they are rapid. The daily feed should be the roughly same amount of the same material each day. Changes in the amount of feed and the material being fed will cause a reduced gas production. Sudden temperature changes will also reduce the gas production, although it requires a great amount of energy to change the temperature of several cubic metres of slurry quickly. Great care should be taken that no chemicals are allowed in the plant that could cause problems. These include antibiotics, detergents, and bleaches. Cleaning materials that contain chlorine will kill the microbes and prevent the plant from working. Small amounts of mild organic soaps can be used for cleaning and should not affect the plant.

Most plants will continue to produce gas for as long as they are fed properly (Kudaravalli, 2014b). If there is a problem with the plant, the methanogens will stop working first. The first indication is that the production of biogas is reduced. This also causes a build up of VFAs, which results in obnoxious smells. The plant is said to go 'acid'. If a plant becomes acidic because of over feeding, it can sometimes be recovered by reducing the feed rate. The addition of a mild alkali, such as calcium or sodium carbonate, can sometimes help.

If gas is not being generated after two or three weeks and the bad smell persists, the only way to recover a plant is to empty it of slurry and start again. If the cause of the problem is the addition of chemicals, the old slurry must be disposed of. Otherwise, it can be stored in a pit and small amounts added to the new feed, once the plant is running effectively again.

If the plant is being fed with animal dung, it should be mixed with water to form a smooth slurry and should be free of lumps. Various tools can be used for mixing the slurry (see Chapter 8). Large amounts of fibrous materials, such as straw and rice husk, will float to the top of the slurry in the plant and can form a mat that prevents gas being released. These should be removed while the slurry is being mixed. Inert materials, such as sand or clay, can be picked up when animal dung is collected. This material will sink to the bottom of the digester and reduce its working volume over time. Mixing pits can be designed that allow these materials to settle out during mixing so they do not enter the main digester. The base of the pit can be sloped away from the entrance to the digester or a ridge made in the bottom close to the inlet pipe, which prevents the bottom layer of the slurry from entering the main pit. The material held behind the ridge, which will contain the silt of mud and sand, can be scooped out and placed on a garden or field.

It is often recommended that biogas plants are emptied after they have been used for between five and ten years, so the silt that builds up can be removed. Few people are willing to make the effort that this requires. However, if gas production does reduce steadily over time, this may be the cause and the plant may need to be emptied.

The slurry can be bailed out, using a bucket on a rope, into a pit that has been dug for the purpose. As much slurry should be removed from the top as possible. There is a danger that the remaining slurry is still producing gas and filling the pit with carbon dioxide and methane, which would suffocate any person who enters. The remaining slurry should be washed through with clean water poured down the inlet pipe and removed using the same bucket and rope. Even when the slurry pit appears to be clean, the danger of suffocation remains. It has been recommended that an air blower be used to flush out the pit (van Buren, 1979). A small animal (such as a chicken) can be lowered into to the pit in a basket to check that the air is breathable. If it survives, a person can get into the pit to clean out the silt.

References

Ashden (2005a) 'Biogas Sector Partnership, Nepal', [website – case study] Ashden <www.ashden.org/winners/BSP> [accessed 21 July 2014].

Ashden (2005b) 'Kigali Institute of Science, Technology and Management (KIST), Rwanda', [website – case study] Ashden <www.ashden.org/winners/kist05> [accessed 21 July 2014].

Ashden (2006) 'Appropriate Rural Technology Institute: Compact digester for producing biogas from food waste', [website – case study] Ashden <www.ashden.org/winners/arti06> [accessed 21 July 2014].

Ashden (2007a) 'Biotech: Management of domestic and municipal waste at source produces biogas for cooking and electricity generation', [website – case study] Ashden <www.ashden.org/winners/biotech> [accessed 21 July 2014].

Ashden (2007b) 'SKG Sangha, India, Biogas for cooking plus fertiliser from slurry', [website – case study] Ashden <www.ashden.org/winners/skgsangha> [accessed 21 July 2014].

Ashden (2010a) 'Ministry of Agriculture & Rural Development and SNV, Vietnam', [website – case study] Ashden <www.ashdenawards.org/winners/MARD10> [accessed 21 July 2014].

Ashden (2010b) 'Sky Link Innovators, Kenya', [website – case study] Ashden <www.ashden.org/winners/Skylink10> [accessed 21 July 2014].

Bulmer, A., Finlay, J., Fulford, D.J. and Lau-Wong, M.M. (1985) *Biogas: Challenges and Experience from Nepal Vols. I and II*, Butwal, Nepal: United Mission to Nepal. <www.kingdombio.com/Biogas-vol-I.pdf> [accessed 21 July 2014].

van Buren, A. (1979) A Chinese Biogas Manual, London: Practical Action Publications (IntermediateTechonology Publicatons). <http://developmentbookshop.com/chinese-biogas-manual-pb> [accessed 22 July 2014].

Cheng, S., Du, X., Xing, J., Lucas, M., Shih, J. and Huba, E.-M. (2011) '4-In-1 Biogas Systems: a Field Study on Sanitation Aspects & Acceptance Issues, in Chaoyang and Shenyang Municipalities, Liaoning Province' Beijing, China: *Centre for Sustainable Environmental Sanitation*, University of Science and Technology, Beijing. <www.ecosanres.org/pdf_files/4-in-1_Household_Biogas_Project_Evaluation-20110620.pdf> [accessed 21 July 2014].

Davis, C.H. and Preston, T.R. (1983) 'A combined digester and gasholder - PVC plastic tube biogas unit', *ADAB NEWS* 10 (January-February): 35–41 <http://biogas.wikispaces.com/Plastic+Bag+Digesters> [accessed 23 July 2014].

Devkota, G. (2003) *Biogas Installation and Training, METTA Development Foundation, The Burma Sustainable Development Project (BSEP)*, Grassroots Leadership Training (GLT), Alam, Myitkyina, Myanmar <www.palangthai.org/en/bsep/docs/BiogasTrainingOctNov2003.pdf> [accessed 21 July 2014].

Devkota, G.P. (2001) *Biogas Technology in Nepal: A sustainable source of energy for rural people*, Kathmandu: Mrs. Bindu Devkota, Maipee.

Fulford, D. (1988) *Running a Biogas Programme: A handbook*, London: Practical Action Publications (Intermediate Technology Publications). <http://developmentbookshop.com/running-a-biogas-programme-pb> [accessed 22 July 2014].

Heeb, F. (2009) *Decentralised Anaerobic Digestion of Market Waste: Case study in Thiruvananthapuram, India*, Dübendorf, Switzerland: Swiss Federal Institute of Aquatic Science and Technology (Eawag). <www.eawag.ch/organisation/abteilungen/sandec/publikationen/publications_swm/downloads_swm/ad_market_waste.pdf> [accessed 21 July 2014].

Hunt, S. (2009) *Small-Scale Bioenergy Initiatives: Brief description and preliminary lessons on livelihood impacts from case studies in Asia, Latin America and Africa*, FAO and PISCES, Rome. <ftp://ftp.fao.org/docrep/fao/011/aj991e/aj991e.pdf> [accessed 21 July 2014].

Idan, J.A. (2008) 'Presentation on Integrated Sewage And Solid Organic Waste-To-Energy Project', [website – presentation] <www.biogasonline.com/downloads/integrated_waste_to_energy_project.pdf> [accessed 23 July 2014].

Khandelwal, K.C. (2008) *Country Report on Financing of Domestic Biogas Plants in India*, SNV., Jaipur, India <www.snvworld.org/en/Documents/India_international_workshop_on_financing_of_domestic_bi ogas_plants_2008.pdf> [accessed 21 July 2014].

Kudaravalli, K.K. (2014a) *Construction Manual for Biogas Technology*, Kolar, South India: SKG Sangha. <http://foundationskgsangha.org/Construction%20Manual.pdf> [accessed 21 July 2014].

Kudaravalli, K.K. (2014b) *Users Manual for Biogas Technology*, Kolar, South India: SKG Sangha <http://foundationskgsangha.org/Users%20Manual.pdf> [accessed 21 July 2014].

Lüer, M. (2010) *Installation Manual for Low-Cost Polyethylene Tube Digesters*, Bolivia: GTZ. <https://energypedia.info/images/1/19/Low_cost_polyethylene_tube_installation.pdf> [accessed 21 July 2014].

Maramba, F.D. (1978) *Biogas and Waste Recycling: The Philippine experience*, Metro Manilla, Philippines: Maya Farms Division, Metro Flour Mills. <www.scribd.com/doc/56576420/Biogas-and-Waste-Recycling> [accessed 23 July 2014].

Nazir, M. (1991) 'Biogas plants construction technology for rural areas', *Bioresource Technology* 35: 283–289 <http://dx.doi.org/10.1016/0960-8524(91)90126-5> [accessed 23 July 2014].

Pal, S.P., Bhatia, V.K., Pal, B. and Routray, D. (2002) *Evaluation Study on National Project on Biogas Development*, New Delhi: Programme Evaluation Organisation, Planning Commission of India. <http://planningcommission.nic.in/reports/peoreport/peoevalu/peo_npbd.pdf> [accessed 21 July 2014].

Satyamoorty, K. (1999) *Biogas - a Boon, Handbook on Biogas*, Kanyakumari, South India: Vivekananda Kendra – NARDEP. <www.vknardep.org/publications/english-books/189-biogas-a-boon.html> [accessed 23 July 2014].

VK NARDEP (1993) *Biogas: a manual on repair and maintenance*, Kanyakumari: Vivekananda Kendra, Natural Resources Development Project. <www.vknardep.org/publications/english-books/197-biogas> [accessed 21 July 2014].

White, R. (2005) *The Role of Biogas in Rural Development and Resource Protection in China: A case study of Lijiang Municipality, Yunnan Province, China*, National Science Foundation and Michigan State University. <http://s3.amazonaws.com/zanran_storage/forestry.msu.edu/ContentPages/16804398.pdf> [accessed 23 July 2014].

Yin, C. (2001) 'Using integrated biogas technology to help poor communities in Baima Snow Mountain Nature Reserve', *Boiling Point* <www.hedon.info/UsingegratedBiogasTechnologyToHelpPoorCommunitiesInBaimaSnowMountainNatureReserve> [accessed 21 July 2014].

CHAPTER 10
Management of a biogas programme

Abstract

Lessons can be learnt from the way biogas programmes have been run in places such as China, India, and Nepal. The way a programme is run depends on the local environment and, particularly, the needs of local people for the technology. One key aspect of a biogas programme is a balance between central and local management. Biogas technology, as a renewable energy, attracts central political interest, which can often help an extension programme. However, the work of selling and building biogas units and appropriate follow-up needs to be done by trained local people. These local extension agents need to be identified and given a range of important skills. Programme finance is an important issue, especially if customers can receive subsidies and loans to enable them to purchase biogas units. Carbon offset finance has been used to provide extra funding, but there are challenges.

Quality control of biogas units is essential and can be linked to the provision of finance. Many programmes need to include research and development work to enable the technology to be kept up to date and to take advantage of new opportunities for its use.

Keywords: biogas, anaerobic digestion, management

Lessons from existing biogas programmes

While many people have built individual biogas plants for their own use, or small numbers for a community, a biogas extension programme works most effectively when a large number of plants is built in an area. A well-organized biogas extension programme allows technicians and other staff to develop good skills and ensures that those skills remain in the area. Different biogas programmes have adopted different management approaches, so it is possible to learn from their successes and failures.

The early biogas programmes in China, India, and Nepal faced two conflicting requirements: they needed to be centrally managed, but the best way to ensure effective extension of a new technology was through local small organizations. In China, the initial emphasis was on local installation, but large numbers of plants failed because of poor quality-control. The solution was increased central control of the programme. In India, the initial emphasis was on central management, which resulted in slow uptake of the technology.

http://dx.doi.org/10.3362/9781780448497.010

This was overcome by allowing local NGOs to become involved in installation and the running of local programmes.

In Nepal, the biogas work in Development and Consulting Services DCS was set up as a small project to support a biogas programme initiated by the Ministry of Agriculture, but then needed to establish the Gobar Gas Company (GGC) to develop the original programme (Fulford, 1988). The GGC had to grow rapidly as it took responsibility for the national biogas programme (Karki *et al.*, 2007). GGC then faced the problems of being a centrally-managed programme that might become top heavy (Fulford *et al.*, 1991). BSP (Biogas Support Programme) was set up when the Netherlands Development Organization (SNV) took over the programme (Karki *et al.*, 2007) and divided the installation work between more than 40 local installers (which increased to 65 and then to 107 over many years). This approach has managed to balance these apparently conflicting requirements. The programme in Nepal can claim more biogas plants per head of population than any other country in the world, while also claiming a 98 per cent success rate based on plants that were still working five years after they were installed (Ashden, 2005a).

A successful programme to extend the use of a novel technology, such as biogas, needs to manage a large number of factors effectively. The programme needs expertise in a wide range of fields, including technology, finance, management, logistics, agricultural extension, and publicity. The people who actually sell, install and service biogas plants need to understand the local culture and environment so they can relate well to the local people. This means that local people should be employed and given good training from the experts. The most effective programmes need both central and local aspects.

More recent biogas programmes have been started in response to specific local needs. The programme in Rwanda (Ashden, 2005b) was designed to meet the need for sanitation in overcrowded prisons. The programme in Ghana (Idan, 2008) has the aim of improving sanitation in institutions such as hospitals, schools, and colleges. The food-waste biogas programmes of Biotech Ltd. and Appropriate Rural Technology Institute ARTI (Ashden, 2007, 2006) in India were set up in response to an increasing demand for cleanliness from a fast-growing middle class. Traditionally in India, cattle that have been allowed to wander freely in towns and villages have consumed food waste. The rapid expansion of many cities has led to an increase in the land value of the small gardens where these animals traditionally lived, so the land has been sold to developers. The number of cattle in Indian cities and towns has dropped sharply making the disposal of food waste a local political issue in many places.

The specific local needs will determine the way a programme is run. However, it is important that the programme plans are not distorted by these needs. Care must be taken in the training of staff and the setting up of appropriate procedures to ensure the programme runs effectively. A real danger for the effectiveness of a programme is to set it up too quickly to meet local short-term needs.

Central coordination and local management

National and local governments supported the three successful programmes in Nepal, India, and China. Each had a coordination committee, composed of representatives from the various organizations involved in the work. In Nepal, this committee had civil servants from the ministries of Agriculture, Energy, and Industry, so that political decisions could be made that allowed the programme to work effectively. Biogas Sector Partnership in Nepal BSP/N is now an NGO that reports to Alternative Energy Promotion Centre (AEPC), a government institution under the Ministry of Science, Technology and Environment.

When SNV extended its programme to Vietnam and Bangladesh, it tried to set up similar coordinating committees (Fulford *et al.*, 2012). The Vietnamese programme is run jointly with the Ministry of Agriculture and Rural Development (MARD), so the government is closely involved. However, in other countries some ministries have proved less willing to be involved in a coordinating committee. For example, in Bangladesh, the programme is run under Infrastructure Development Company Limited (IDCOL), which was set up as a government corporation. There is a coordinating committee at government level, but some government ministries give poor support.

A coordinating committee needs to be set up fairly early on in the programme, otherwise problems can occur as the people running the programme seek to obtain permissions and licenses for their various activities from different government departments. Along with civil servants, the committee should include representatives from the banks and other finance groups involved in loan finance, as well as senior people involved in manufacturing, extension, and research and development for the programme (van Nes *et al.*, 2009). A coordinating committee allows any policy conflicts and other problems to be sorted out at a high level.

It is also helpful to set up similar committees at the local programme level, to ensure that local organizations are working well together. Again all those involved in the programme (at regional, state or provincial level) should be represented, along with bank representatives, local government officials and politicians. This approach has worked particularly well in India, where local government officials have been very supportive of different biogas programmes.

SNV has a policy of involving multiple stakeholders in the work of planning a new biogas programme in a country (van Nes *et al.*, 2009). People from government departments, private companies, finance organizations and NGOs all have a part to play in the programme. SNV suggests that each group should only be involved in the areas for which they have existing expertise. This means the functions required to develop a programme need to be shared between the different organizations involved.

Biogas programmes have not functioned as well in Africa and have not achieved the numbers of plants seen in some countries in Asia. The main reason is that national governments have not given their whole-hearted support. Most of the programmes have been started by local enthusiasts or outside aid organizations. Without good support from government officials and particularly from banks and other finance institutions, these programmes have struggled to be effective. Some African countries have allocated the responsibility for renewable energy to a government department. Confusion can occur when various departments are given responsibility for different aspects of a programme. Biogas technology can come under departments such as Agriculture, Environment, and Energy, all of whom have an interest.

A central managing committee can cause problems in the actual running of the programme, if it attempts to control the programme too closely. The programme in India was run under the National Project on Biogas Development (NPBD), which defined central and local targets for the construction of biogas plants. This approach often meant that plants were built for people who were not ready to use them. It also meant that extension workers were not able to take advantage of new opportunities because these not been anticipated in the yearly plan. In Bangladesh, the Bangladesh Council for Scientific and Industrial Research (BCSIR) had a tight control of biogas development, but saw it in terms of research, rather than an aspect of rural development. SNV decided to work with another government agency, IDCOL (Infrastructure Development Company) because they wanted to use a different approach (Fulford *et al.*, 2012).

The most effective people to sell and build biogas plants in a local area are people from that area. Small local extension groups can relate to local people and understand the most effective way to work with them (Leermakers, 1992). However, there needs to be a means to relate these local groups with the national committees, so they can be part of the national programme.

In India, NPBD became much more effective, when it allowed local initiatives to be more involved and to use their own approaches. For example, the Planning Research and Action Division of the Uttar Pradesh State government (PRAD) and NGOs, such as Action for Food Production (AFPRO) were able to train and manage local biogas initiatives, which allowed the programme to develop faster.

In Nepal, the biogas plant installation programme was set up as a private company Gobar Gas tatha Krishi Yantra Bikash Pvt Ltd (Biogas and Agricultural Equipment Development Private Ltd) known as the Gobar Gas Company (GGC). This was done so that the Agricultural Development Bank of Nepal (ADB/N) and United Mission to Nepal (UMN) could work together in the programme. The Nepal Fuel Corporation was another shareholder. GGC had several local offices and depots across Nepal, which related closely to the local banks set up by ADB/N.

After UMN (an aid organization) had handed over its interest to ADB/N (a bank), GGC continued to grow until it became too large and unwieldy as an organization. By 1990, the programme had expanded further, mainly as a result of subsidies being made available through a finance package offered by the United Nations Capital development Fund (UNCDF). The local office personnel were losing motivation and quality standards were beginning to drop. The success rate for biogas plants had dropped to about 90 per cent.

In order to overcome these problems, the Biogas Support Programme (BSP) was set up in 1992 (Devkota, 2001). The plan was to license a number of biogas installation companies, each of which would work in a local area, under a central organization. The task of BSP was to provide training, quality control and technical back up from people involved in research and development. A number of local GGC managers left the main company and formed their own construction companies. Other small construction companies saw the opportunities that the programme offered and wanted to join in. Initially about six new biogas companies became part of BSP along with GGC and the number grew rapidly to 49 (Karki *et al.*, 2007). The numbers have varied slightly, but by 2011 the programme has grown to 82 members (Fulford *et al.*, 2012).

The BSP approach, with local installation companies under central coordination, has worked very well, with a rapid expansion in the plants sold to customers. BSP is involved in effective quality control of the plants that are being built, so that installers have to meet high standards. This involves visits to new plants by BSP personnel, as well as follow-up visits to make sure the plants continue to work (Lam, 1996). The success rate for plants in Nepal went back up to 98 per cent (plants that were still working five years after they were installed). BSP had built some 250,476 plants by the end of 2011 (Energy4All, 2012).

However, the key to making this process work is that the installer can only recover the cost of the plant from ADB/N, for both subsidy and loan components, if the plant passes the quality-control standards. Any errors in manufacture must be put right before payment is made. This gives a very high incentive for the manufacturers to ensure all their staff are trained to the highest standard.

In China, the Chengdu Biogas Scientific Research Institute of the Ministry of Agriculture seems to perform a similar role to that of BSP in Nepal. It sets standards for biogas technology and trains technicians in those standards. The Institute does not appear to have such a tight control over biogas plant construction as BSP in Nepal; in China there are independent groups involved in the dissemination of biogas technology. However, the poor quality of the plants that were built in the decentralized programme in the 1970s persuaded people in the government that a centre for biogas standards and training was essential. By the end of 2011, China claimed to have built 42.8 million biogas plants (Energy4All, 2012).

The approach in India seems to have been the opposite of that in China (Moulik, 1985). After a period of fairly tight central control under Khadi and Village Industries Commission (KVIC), the involvement of local groups and NGOs allowed a relaxation of control over the technology. While groups such as PRAD and AFPRO did have high technical standards and offer high-quality training, a few NGOs and small commercial concerns were less careful, which led to the building of poor-quality biogas plants in some areas of India. The government, through its Ministry of Non-conventional Energy Sources (MNES) is offering subsidies to installation groups, which gives an opportunity to exercise a higher level of quality control. It appears that some local government groups are less careful than others. In many biogas programmes, there are very weak levels of quality control, which has led to a high level of failures. The best Indian biogas extension programmes, such as SKG Sangha, use internal quality-control procedures. The number of plants claimed to have been built in India was 4.4 million (Energy4All, 2012).

The move from subsidies offered by governments to that from carbon offset finance brings a new opportunity for groups to offer quality control of the plants that are being built. The Clean Development Mechanism (CDM) does insist on very detailed methodologies (CDM, 2005) that must be listed and followed. These methodologies depend on plants saving carbon dioxide over several years, so the implication is that the technology must be reliable over that period of time. The methodology will usually define an explicit process by which this quality control can be ensured.

Groups involved in the voluntary carbon offset market also recognise the need to include similar quality-control measures in the agreements that they make with biogas programmes. Again, the agreement will include guarantees that the technologies that save carbon dioxide, such as biogas plants, will be reliable and continue to work for the length of the agreement. In order for this to happen, the installations need to be of a quality that ensures the plants do not fail before the end of the agreement. Since both CDM and voluntary groups involved in carbon offset agreements require sufficient carbon dioxide savings to make the agreements worthwhile, the number of biogas plants that will be included in such an agreement will be fairly large. There is a desire to aggregate smaller projects into groups to enable sufficient numbers of plants to be considered in such an agreement. This approach will encourage a centralized approach, which will allow organizations to monitor the quality of the work of the smaller projects that want to become involved.

Biogas extension work

The actual job of selling and installing biogas plants is done by local people acting as extension agents. These agents can be employed directly by the central programme, but if extension is to be done by local groups, they can be employed by a local NGO or a small construction company.

The extension agent usually has to perform a wide range of tasks (Fulford, 1988). The tasks include selling: persuading local people to have a plant installed and working with the local bank to obtain loans and a suitable subsidy, if one is available, so that the customer can pay for the plant. The next phase is to arrange for building materials to be delivered to the site and to ensure that construction workers are available to build the plant, as soon as the materials are delivered. Often the same extension agent acts as a construction supervisor, to make sure that the plant is built to the correct standards. Agents work with their customers to ensure that they understand the technology and use it correctly and that they do not have any problems. This follow-up work is often a condition of the loan because banks want to make sure that the plants continue to work well, so they get their loan repayments.

The logisitics of ensuring the correct amounts of each item of the required building materials are one site when they are needed are complex. If an item is missing, the construction workers may need to wait until it is available. In more remote areas, such as in many parts of Nepal, the logistics of transporting materials to a building site is even more complex. Some materials, such as aggregate and sand can be sourced locally, but cement and metal parts may need to be carried by people from the nearest road.

This suggests that the local extension agents need to be well trained in the technology of making biogas plants, in effective selling techniques and in the management of complex logistics (Karki *et al.*, 1996). Agents need to know how to look for potential customers and how to convince them of the value of a plant. Western approaches to hard selling are unlikely to be effective in more remote areas; the sales techniques need to be culturally appropriate for the locality. The most effective agents tend to come from the same area and understand the local cultures. Young people in most places are now getting better education than in previous years, agents must have basic writing and arithmetic skills because they need to fill in forms and do basic calculations. Finding suitable local people to train, who have the right skills, is much less difficult than it used to be.

The extension agents do need a thorough basic understanding of the technology of biogas, how it works, and how the plants are made. When working with the customers to sell a unit, they need to know how to choose the correct size of plant, based on the amount of feedstock available and the gas the customer needs. They also need a reasonable understanding of quantity surveying, to make sure they have the correct quantities of materials for the size of plant they are building. The customer is often required to supply some of the materials, such as sand and aggregate, and also labour to dig the digester pit and to mix the concrete. The agent needs to make sure all these things are done correctly.

The agents need to be able to explain to customers what they are buying and how to use it properly. They need to quickly identify errors that the construction workers might make, so that they can maintain good quality-control. They need to train the customers in the correct ways to add feedstock to the plant, how to use the gas safely and economically, and how to do simple maintenance, such as cleaning gas stoves and clearing water traps.

These agents also need to work closely with the local bank officials, so they can help the customers fill in the right forms to obtain the correct loans and subsidies. The price the customer pays for the plant will relate to how much they can offer in materials and labour towards the project. The bank will do these calculations, but the agent needs to make sure that the correct items are included so that the final price paid by the customer is right.

Travel in remote areas can add greatly to the cost of the programme. Some programmes expect extension agents to use public transport and bicycles, but while the cost of travel is much reduced, the productivity of the workers is also much reduced. Land cruisers can negotiate poor roads, but are not usually economic if all the extension agents need to use them. In Nepal and India, the most effective approach seems to be the provision of motorbikes. The cost of running and maintenance is significant, but the amount of extra work the extension agents can achieve in a given time justifies the cost. In hilly areas, where roads are limited, as in many areas of Nepal, there is no real alternative to walking. It was found that horses and mules were too expensive to feed and look after.

The construction side of the programme usually involves a small team of people, who work with the extension agent. The team is usually led by a master mason with one or two assistants. If the customer does not provide the unskilled labour, the team will find people to employ, often hiring local people for the work. The team might include a specialist pipefitter to connect the gas lines. The team might also include a supervisor, if the extension agent is not able to fulfil this role. When GGC ran the Nepal biogas programme the area offices employed the teams.. However, the use of small independent contractors has proved much more successful. Each contractor employs one or more teams to build plants in specific local areas. The members of the teams are trained by BSP to a high standard (Gurung, 1996) by working with master masons who have demonstrated their skills. The quality of a randomly selected number of plants built by each team is inspected by BSP before the bank makes payment to the company.

Small companies are usually better than large contractors because their construction teams are more flexible and can adapt to the needs of the programme. Larger contractors do not have the local contacts that smaller firms will have. Also larger companies can be affected by political influences, which can conflict with the aims of the programme. In Nepal, the Gobar Gas Company continued to operate as one of the many companies building biogas plant. However, many of the area managers left to set up their own companies, based in the area in which they had built their contacts.

One issue faced by small construction companies in both Nepal and Bangladesh has been the retention of masons. The construction companies and wider project provide good training to masons to enable them to build quality biogas systems. A monsoon climate means that construction cannot take place in the wet season, so masons do not have work for two or three months a year. Many masons look elsewhere for places where their new skills can be used and they find better paid work, often in other countries, such as the Middle East (Fulford et al., 2012). This means that each year, a new group of masons need to be trained, so that the work can proceed.

In general, the best groups to do biogas extension work are local contractors. These contractors can be either non-profit NGOs, or for-profit companies. Some of these groups may be larger than others. One of the keys to the effectiveness of the programme in many countries is central monitoring, but the work must be done properly.

Programme organization

The middle management of a biogas programme acts as the link between the national or regional coordinating committee and the local construction teams or companies. The organizations that run the programme can be government, NGO, or commercial. The Gobar Gas Company in Nepal was set up as a commercial organization. The costs of administration, as well as plant construction, were added to the charges to customers, so this approach did make the technology expensive. BSP was funded by SNV, which removed the costs of central administration from the actual extension work. In China, government groups run the main programme, particularly the Chengdu Biogas Research Institute, which defines national standards and offers training, although the actual construction work is done by small NGOs or local companies. In India, the programme is run by a mixture of government organizations, such as KVIC local government groups, such as PRAD and NGOs, such as AFPRO. However, independent groups, such as SKG Sangha, have done the most effective extension work with little involvement from central organizations.

The management task is to provide support for the extension groups, especially quality control, training, financial coordination, manufacturing, material supply, publicity, and research and development. Appropriate functional managers will be responsible for these different tasks and will work together to make sure that the tasks are coordinated. One of the main tasks of the group is to make yearly plans for the progress and expansion of the programme, which can be submitted to the central coordinating committee. These plans will include a budget to finance the programme, the amount of loan finance required from the banks as well as the amount of subsidy that will be needed. These plans rely on data obtained from the extension groups for the expansion potential in different local areas.

Where a programme is government or aid funded, plans are usually made for blocks of a few years at a time. In China, India, and Nepal, it has been common to run a programme in various phases, as it is easier to obtain commitments from funding bodies. For example, Phase IV of the BSP programme ran from July 2003 to June 2009 (Fulford *et al.*, 2012). Such a plan looks at the number of plants to be built each year, the resources required to meet the targets and the cost of those resources. A similar approach is being used by SNV in other countries where they have developed similar programmes, such as Bangladesh and Vietnam (van Nes, 2007). The cost of these programmes is high, so budgets and sources of finance need to be planned very carefully. If finance is required from outside the programme itself, proposals need to be written that give all the details of the programme. These proposals need to be presented to funding organizations, such as UN agencies or government aid organizations.

Finance is required not only for the operation of the programme, but also for loans and subsidies that allow customers to purchase biogas plants; the most effective organization to handle money is a bank. In many Asian countries, there are rural banks that specialize in lending to small farmers. In Nepal, the ADB/N was highly involved in the biogas programme from its inception. Other banks were drawn into the programme, as it grew and required more financial input. In India and Bangladesh, the Grameen banks have been set up to provide loans for rural farming activities. These banks have taken an interest in biogas technology and have enabled the programmes to develop. Grameen banks have the advantage that they have offices in villages, so farmers do not have to travel long distances to get financial help. The managers and staff become well known to the local people, so people begin to trust them and feel confident in taking out loans.

The situation in China has been somewhat different in that the government offered its subsidy in the form of materials, while labour was seen as the duty of the local community. Individuals received help to build biogas plants from the local community, but had to be willing to help build plants for other people. As the Chinese economy has become less communal, banks are becoming more involved, but the government still has strong control over the programme.

One task of the central organization is the purchase of materials in bulk for the programme. The material that is required in large quantities is cement. Some designs also need quantities of bricks. Biogas plants also need supplies of steel reinforcement and gas pipes. These items can then be distributed to the local contractors against future payments for the work they will do. Gas equipment, such as stoves and lights, is also required. These may be purchased from a supplier, or manufactured by other contractors in the programme. If the programme is responsible for the manufacture of such items, they can offer the same training and quality-control oversight to contractors as they provide for plant installers. Since the central organization has clear plans for the programme, it can estimate the quantity of different materials and items required and ensure these materials are provided in the right place at the right time.

Programme finance

Biogas programmes need finance both to launch and run the work and also to assist customers to purchase units. Government departments initially launched the programmes in China, India and Nepal, so programme finance was not an immediate issue. However, there was a need to provide loans to farmers to purchase plants. As with most renewable energy technologies, biogas has a high initial cost, but saves money as it is being used. A loan allows a customer to cover the cost and the savings allow them to repay the loan.

The Indian government has been involved in subsidizing the building of biogas plants since 1981 and the programme is still continuing (Khandelwal, 2008). The emphasis has been on the achievement of targets, rather than the quality of the plant manufacture. There is a national subsidy administered by the MNRE (Ministry of New and Renewable Energy) through the NBMMP (National Biogas and Manure Management Programme) and also local subsidies from State and District governments. The subsidy is set at a fixed value each year, rather than as a percentage of the cost. This means that a farmer having a small plant receives a higher percentage of the total cost than someone who has a larger plant. However, the value of the subsidy has not been increased in line with the rising costs of building plants.

In India, the banking system is very much involved in providing loans and subsidies, especially the Reserve Bank of India (RBI) and the National Bank for Agriculture and Rural Development (NABARD). The Indian government has always worked through national and local banks to provide finance for development projects. However, the banking system in India has always been highly centralized and bureaucratic, so the process of obtaining loans and subsidies has been complex. The biogas programmes have required staff who can liaise between the local farmers and the bank officials to process all the paperwork required. In more recent years, the rise of micro-finance institutions (MFIs) has enabled the provision of loans to the rural poor to be simplified and streamlined.

In Nepal, the Agricultural Development Bank of Nepal was very involved in the biogas programme from the beginning. Representatives from ADB/N were members of the Board of Directors of the Gobar Gas Company, before the formation of BSP Nepal, so they were keen to administer loans and subsidies to customers through their local offices (Bajgain and Shakya, 2005). BSP was established to allow a much larger number of groups to be involved in biogas construction. This more open programme was also able to draw in other banks and MFIs to be involved in the provision of loans (Dhakal, 2008).

The early biogas programmes in China were subsidized, but in-kind, rather than with cash. People who wanted to build a biogas plant were provided with free labour and sometimes with extra land. The people involved in making biogas plants had little experience or technical training and there was little monitoring of the quality of the work. In 2003, the government established the National Loan Subsidy Programme (NLSP) in China, which provides a cash subsidy for biogas plant construction (Zhang, 2008). Aid organisations, such as the Japanese Government, have added money into this fund to encourage its growth. Some Chinese provincial governments have added an extra subsidy to the national one. Farmers can get up to 50 per cent subsidy from NLSP and another 20 per cent to 30 per cent from local government to build their plant. These subsidies have encouraged a rapid expansion in the building of biogas plants in China.

In more recent years, carbon offset finance has been seen to be a useful source of finance for biogas extension programmes (Capoor and Ambrosi,

2007; ter Heedge, 2008). BSP in Nepal pioneered a methodology for CDM in 2005, which was initially successful. A change of policy meant that this approach could not be repeated, so a completely new methodology had to be written, which eventually proved successful. Biogas programmes in Vietnam, Cambodia, and Bangladesh have applied to organizations in the Voluntary Emission Reduction (VER) market, which use similar protocols to the Certified Emission Reduction (CER) market under CDM. The way that such protocols are written have become well defined and also broadened to include Programs of Activity, which can aggregate various projects under a single agreement (Hayashi *et al.*, 2010).

As an NGO running a biogas extension programme in India, SKG Sangha has been very successful in both CER and VER markets, obtaining support for the building of biogas plants for domestic use. Unfortunately the value of carbon offsets reduced between 2006 (almost $42 a tonne) (Capoor and Ambrosi, 2007) and 2012 (below $9 a tonne) (Peters-Stanley and Yin, 2013), due to the political situation. The Kyoto Protocol, an agreement between many countries that enabled the CDM to be set up, was not replaced at a meeting of the United Nations Framework Convention on Climate Change in Durban in 2009 (IESE Business School, 2013). A European Carbon Offset Market that was set up to encourage the reduction in the use of fossil fuels in the EU steadily collapsed due to a lack of political support. However, the popularity of the carbon offset market has increased (Peters-Stanley and Daphne Yin, 2013), with the involvement of new players, such as China (IESE Business School, 2013), so this source of finance is likely to continue to be available.

A source of finance available to commercial organizations in a biogas extension programme is that of social investment. Many people are willing to invest finance in projects for a return that is below market rates, if the business in which the investment is made can be shown to provide social benefits. Some national governments provide tax advantages to encourage such investments (HM Government , UK, 2011; Voorhes *et al.*, 2012). However, the usual commercial investment constraints apply: the biogas extension business must be seen to be profitable and the investors require a strong influence over the company, as a requirement for the investment.

Quality control

Financial institutions, such as banks and MFIs, expect customers to repay loans, so the technology that is purchased must be reliable. One of the main tasks of BSP in Nepal was to ensure that all the groups involved in making biogas plants delivered high-quality work. Staff of the construction companies were given training by working alongside people who had a long experience of building such plants. Since BSP was responsible for the administration of subsidies and loans for these plants, it was able to monitor the quality of construction of the plants. Constructors were only paid for their work if it met the required standard.

The high standards achieved for biogas plants in Nepal was seen as one of the reasons for the growth of the project (Karki *et al.*, 1996). People felt they could trust BSP to supply them with a working plant. If there were any faults, the installers were motivated to put them right quickly, to ensure they would receive payment. BSP's quality-control process – with a list of 66 parameters to check – had become well established with time (Karki and Ghimire, 1998). BSP started by sending experienced technicians to check every plant that was built. As the programme developed, BSP randomly selected a sample of plants installed by each company that was licensed to build plants. The 66 parameters were placed in three categories: category 1 errors meant that the plant would not receive a subsidy; category 2 errors resulted in a penalty (i.e. part of the subsidy was withheld); category 3 errors meant the constructor was given a warning. The assessors assumed that any error found in the sample occurred in all of the plants built by that company, so the penalty was charged for every plant built by that company that year.

BSP has defined detailed plans for each plant built under their programme. Originally, the approved design was the one developed by DCS for GGC. The design was simplified and updated in 1992 and called the GGC 2047 model. A version of the Deenbandhu plant, as developed in India, was added to the approved list in 1995. All plant dimensions had to be within the tolerances defined by these plans. The technicians employed by each company had to have attended a two-month training course on building biogas plants approved by BSP. The quality of the materials used in the biogas plant, including bricks, sand, gravel, and cement, also need to be within defined specifications.

The success of the BSP project encouraged SNV to use a similar approach in its programmes in other countries, such as Bangladesh (van Nes, 2007), Vietnam, Rwanda, and elsewhere. However, in local programmes in countries other than Nepal, the quality-control culture seems to be less well established (for example in Vietnam (SNV, 2006)).

The attitude to biogas technology in China changed once a set of standards had been established in 1985 and construction groups were encouraged to use them (van Nes *et al.*, 2009). However, because of the lack of quality standards in the early 1980s, it has taken many years for people to overcome their initial impression that biogas technology failed within a few years. The introduction of the government subsidy scheme meant that quality standards could be imposed – constructors have to meet them before the subsidy is paid on the plant.

In India, the government has placed emphasis on the achievement of targets for the number of plants built, rather than the quality of the construction of the plants. This has meant that some groups constructing plants have cut corners to achieve numbers and the quality of the plants has sometimes been poor. A study in 2001 (Khandelwal, 2008) of 615 plants that were supposed to have been built, found that 81 per cent had not been finished. Of those that had been commissioned, only 61 per cent were actually working. The lack of quality control and effective follow-up has given biogas technology a poor reputation in some areas of India.

SKG Sangha stands out as an organization doing biogas extension in India that has a very effective quality-control process. Its success rate is very high for two main reasons, the people who do the construction work are usually local to the area and there is a high emphasis on quality control and follow-up. If customers do have issues with the running of their plants, they know the people who did the work and can contact them directly. They are also given a telephone number for the main office of the extension organization so they can contact the managers if they have a problem.

Employment issues

Most biogas extension programmes employ people in two ways. Some people are permanent employees with contracts. These people are mainly the administrative staff, the engineers, and senior supervisors. These people are paid a fixed salary, but may also receive incentive payments, depending on their jobs. Staff do need training to develop their skills and should have the opportunity to advance through the organization, if they have proved their capabilities.

Many of the construction workers, such as masons and junior supervisors, are employed on a casual basis. The construction work is often seasonal. Much of Asia has a monsoon climate, so the work of constructing biogas plants cannot be done during the rainy season. Also, many construction workers prefer to work part-time, because they also have land on which they can grow crops during the monsoon. There is an advantage in employing construction workers from the locality in which a programme is running: once they have been trained to do the quality construction work required for biogas extension, they are in high demand for other construction jobs. When the construction work is finished, they prefer to remain in their locality and do other construction jobs.

However, there is always an issue at the beginning of each construction season that more construction workers need to be trained, as many of the previous workers are doing other jobs. Construction workers from Asia are in demand in other countries, such as the Middle East. The need to train new staff each season is one that should be included in the programme plan. [Fulford *et al.*, 2012]

In Vietnam, the issue is slightly different, as many of the construction workers are self-employed. Once they have received training, they can sell and build biogas plants by themselves, providing they register each plant with the biogas programme administration. They can sell and build plants without registration, but they do not then receive a subsidy. These people are more committed to the programme [Fulford *et al.*, 2012]. However, they can only work within their local area and will do work apart from the biogas programme, if it provides better payment.

Materials management

Once a clear design is available for each size of biogas plant, the amount of materials that are required is clearly defined. These will include bricks, cement, sand, aggregate, and steel reinforcement, as well as gas valves, gas pipes, gas stoves, and other equipment. Some of these materials must be obtained locally, to control transport costs, but other equipment must be purchased from specialist suppliers.

For each plant the materials need to be available before building can start. The customers are often in the best position to obtain local materials, such as sand and aggregate. The customers will usually supply unskilled labour, for example for the digging of the main digester pit. However, the construction team must check the quality of the materials that each customer has provided. The type of aggregate that is required for good concrete, for example, should be crushed rock, rather than the type of pebbles that can be found in river beds, which are too smooth.

The extension group is well placed to purchase other materials, such as gas pipes and gas equipment. It has the capability to purchase these more specialist items in bulk at a good price. The group therefore needs the capacity to store such items, often in localities that are central to the areas in which it is running the building work.

Items such as bricks and cement may be purchased by either the customers or the extension group, depending on whether they are easily available in the locality in which the extension programme is running: Again, there is a need to ensure that these materials are of the appropriate quality. Cement should be as fresh as possible – it weakens with age because it slowly absorbs moisture from the air.

The need to ensure that the correct amounts of all the required materials are available on the building sites is especially important in more remote areas, where transport is more difficult. If some of the members of the team need to travel to obtain missing supplies, the productivity of the whole team is reduced.

Record keeping

Effective quality control and follow-up work requires that detailed records are kept of each plant that is built, the identity of the technicians and supervisors involved in the building of the plant and the number of times that a plant is visited, once it has been built. The requirement from carbon-offset finance organizations for detailed audits of the plants that have been built, and whether they continue to work (i.e. continue to save fossil carbon), strengthens the need for regular and accurate data collection and monitoring.

One issue in many developing areas is that it is difficult to define clearly where people live. The work of census takers and the need for people to have identity cards have meant that addresses are defined, but individual locations can be difficult to find on a map. However, the availability of portable GPS tools that can identify the location of each biogas plant uniquely has made record keeping much more accurate and reliable.

There is a need in many organizations to keep both paper and electronic copies of the database that identifies each plant and the name of the customer for whom the plant was built. Customers need to sign paper copies of agreements, especially if they obtained a loan for part of the cost and also received a subsidy in their name. These agreements need to be filed carefully, so they can be easily retrieved in case of a dispute. The use of electronic organizers with GPS receivers enables records to be made that can be directly downloaded to an electronic database in the main offices of an extension organization. It is important that each plant has a unique identification number, so the paper and electronic records can be easily reconciled.

Good record keeping is also important for materials' management. A good stock-keeping system is needed to ensure the quantity of materials in each store is clearly known and when new stocks should be purchased.

Research and development

The technologies that have proved so successful in the programmes in China, India and Nepal were based on many years of research and development that preceded the start of the extension programmes. Groups in both India and China had been involved in research work for many years, although much of this work went unrecognized until more recently. In China, the early research was done by Luo Guorui (Kangmin and Ho, 2006) and in India, J.J. Patel in KVIC and Ram Bux Singh in PRAD, were the key names (NIIR, 2004). The biogas programme in Nepal benefited from this work, although the initial extension programme was strongly supported by research and development work in DCS, funded by USAID (Fulford, 1988).

Once effective designs of biogas plant had been developed that could be built in large numbers, the work of research and development was dropped. The emphasis in the 1980s was on rural, domestic biogas systems that used animal dung as the main feed material. The work in China was supported by the Biogas Institute of the Ministry of Agriculture (BIOMA), based in Chengdu. BIOMA's main work was setting technical standards for biogas plants and biogas equipment. The research groups in India, such as KVIC, PRAD and AFPRO, turned their attention to other projects. BSP in Nepal chose to remove research and development from its list of activities, although it was later replaced.

The weakness in abandoning research and development was that the main programmes were unable to respond to new opportunities; such as a desire for larger biogas systems and the need to process food waste. Once a technology had been developed that was considered suitable for large-scale extension, it was standardized. No further improvements were considered necessary. In

Nepal, a biogas burner had been developed by DCS in the late 1970s that was much cheaper to make and supply than devices imported from India. The design had not been optimized and a more scientific analysis had shown possible improvements. These were not incorporated because the burner was already in production. A lack of emphasis on research and development meant that this burner design was not properly tested until 2004 (CES, 2004). Even then, improvements were not made and similar designs were exported to other places in the world, such as East Africa (Tumwesige *et al.*, 2014).

The development of biogas systems to meet new opportunities has been done by small groups in various places. The development of food-waste digesters for small-scale domestic use was started by Dr A. Karve of ARTI (Ashden, 2006), using a floating drum design made from HDPE water tanks. Biotech Ltd., another group, developed a similar system (Ashden, 2007), using a floating drum design made from fibre reinforced cement (FRP). Larger two-stage food-waste digesters were also developed by Biotech Ltd., as well as Dr S.P. Kale in Bhabha Atomic Research Centre, who developed the Nisargruna process (Shah, 2006). Both of these systems use KVIC-design floating drum digesters, so are seen as expensive. Biotech has recently developed a modular two-stage digestion system made from FRP, although the cost remains high. It is also testing tent digesters.

The use of large size Deenbandhu designs was developed in Africa. Ainea Kimaro, an engineer from Tanzania, built fixed-dome digesters in Rwanda (Ashden, 2005b) with internal volumes of 100 m³ and linked them together to provide systems with volume of up to 1,400 m³. These systems were mainly used to process sewage from jails. Similar fixed-dome plants were built by John Idan of Biogas Technology Africa Ltd in Ghana (Idan, 2008) and Samwel Kinoti of Sky Link Innovators in Kenya (Ashden, 2010). In India, SKG Sangha is introducing building fixed-dome plants with large volumes. The cost of such plants appears to be much lower than for the large floating drum plants that are more widely used.

Much more research and development work needs to be done to optimize the use of biogas technology to process biomass materials other than animal dung. The commercial processing of fruit and vegetables generates various types of residues, but tests are required to discover the most cost effective way to process each material. Many commercial agricultural operations have a requirement for energy and high-quality compost, so there is a good potential for the results of such tests. Examples for energy demand include the processing of mangos to make dried fruit and the removal of coffee berry pulp so that coffee beans can be dried.

References

Ashden (2006) 'Appropriate Rural Technology Institute: Compact digester for producing biogas from food waste', [website – case study] Ashden < www.ashden.org/winners/arti06> [accessed: 21 July 2014].

Ashden (2005a) 'Biogas Sector Partnership, Nepal', [website – case study] Ashden <www.ashden.org/winners/BSP > [accessed 21 July 2014].

Ashden (2007) 'Biotech: Management of domestic and municipal waste at source produces biogas for cooking and electricity generation', [website – case study] Ashden <www.ashden.org/winners/biotech> [accessed 21 July 2014].

Ashden (2010) 'Case study summary: Sky Link Innovators, Kenya', [website – case study] Ashden <www.ashden.org/winners/Skylink10> [accessed 21 July 2014].

Ashden (2005b) 'Kigali Institute of Science, Technology and Management (KIST), Rwanda', [website – case study] Ashden <www.ashden.org/winners/kist05> [accessed 21 July 2014].

Bajgain, S. and Shakya, I. (2005) *Biogas Nepal - Successful model of public private partnership*, SNV Netherlands Development Organisation and BSP, Kathmandu, Nepal.<www.snvworld.org/files/publications/biogas_nepal_-_successful_model_of_public_private_partnership.pdf> [accessed 21 July 2014].

Capoor, K. and Ambrosi, P. (2007) *State and Trends of the Carbon Market 2007*, World Bank, Washington DC. <https://wbcarbonfinance.org/docs/Carbon_Trends_2007-_FINAL_-_May_2.pdf > [accessed 21 July 2014].

CDM (2005) *Biogas Support Program - Nepal (BSP-Nepal) Activity 1*, Board, CDM – Executive: UNFCCC. <http://cdm.unfccc.int/UserManagement/FileStorage/A4NYD8EXQY928HD61LHWHEIM82M BIN> [accessed 21 July 2014].

CES (2004) *Efficiency Measurement of Biogas Stoves*, Pulchowk, Lalitpur, Nepal: Institute of Engineering, Tribhuvan University. <www.snvworld.org/en/download/publications/efficiency_measurement_biogas_stoves_nepal_2004.pdf> [accessed 21 July 2014].

Devkota, G.P. (2001) *Biogas Technology in Nepal: A sustainable source of energy for rural people*, Kathmandu: Mrs. Bindu Devkota, Maipee.

Dhakal, N.H. (2008) *Financing of domestic biogas plants (Nepal)*, SNV Netherlands Development Organisation, Kathmandu, Nepal. <www.snvworld.org/en/download/publications/nepal_financing_domestic_biogas_plants_2008.pdf > [accessed 21 July 2014].

Energy4All (2012) *Brief Progress and Planning Report the Working Group on Domestic Biogas*, ADB, Mandaluyong City, Philippines. <www.hedon.info/tiki-download_item_attachment.php?attId=443> [accessed: 21 July 2014].

Fulford, D. (1988) *Running a Biogas Programme: A handbook*, London: Practical Action Publications (Intermediate Technology Publications). <http://developmentbookshop.com/running-a-biogas-programme-pb> [accessed 22 July 2014].

Fulford, D., Devkota, G.P. and Afful, K. (2012) *Evaluation of Capacity Building in Nepal and Asia Biogas Programme*, Hanoi, Vietnam: Kingdom Bioenergy Ltd for SNV. <www.kingdombio.com/Final Report - whole.pdf> [accessed 21 July 2014].

Fulford, D., Poudal, T.R. and Roque, J. (1991) *Evaluation of On-going Project: Financing and construction of biogas plants*, United Nations Capital Development Fund, New York.

Gurung, B. (1996) *Training Report of Slurry Extension Officers*, BSP/N. SNV, Kathmandu Nepal. <www.snvworld.org/en/download/publications/final_training_report_of_slurry_extension_officer s_nepal_1996.pdf> [accessed 21 July 2014].

Hayashi, D., Michaelowa, A., Dransfeld, B., Niemann, M., Andre Marr, M., Muller, N., Wehner, S., Krey, M. (2010) *Programme of Activities Blueprint Book: Guidebook for PoA coordinators under CDM/JI* (2nd edn), Frankfurt: KfW Bankengruppe. <www.snvworld.org/en/download/publications/poa_blueprint_book_guidebook_for_poa_coordinators_under_cdm_ji_edition_2_2010.pdf> [accessed 21 July 2014].

ter Heedge, F. (2008) Domestic Biogas Projects and Carbon Revenue: A strategy towards sustainability?, SNV Netherlands Development Organisation, The Hague, Netherlands. <www.snvworld.org/download/publications/domestic_biogas_projects_and_carbon_revenue_2008.pdf> [accessed 21 July 2014].

HM Government, UK (2011) *Growing the Social Investment Market: A vision and strategy*, London: HM Government , UK. <www.gov.uk/government/uploads/system/uploads/attachment_data/file/61185/404970_SocialInvestmentMarket_acc.pdf> [accessed 21 July 2014].

Idan, J.A. (2006) *Presentation on Integrated Sewage and Solid Organic Waste-to-Energy Project,* [Presentation] BTAL (Biogas |Technology Africa Ltd) Accra Ghana. <www.biogasonline.com/downloads/integrated_waste_to_energy_project.pdf> [accessed: 23 July 2014].

IESE Business School (2013) *The Future of Global Carbon Markets - The prospect of an international agreement and its impact on business*, Ernst and Young, London. <www.forest-trends.org/documents/index.php?pubID=3898> [accessed 21 July 2014].

Kangmin, L. and Ho, M.-W. (2006) 'Biogas China', *Institute of Science in Society*, London <www.i-sis.org.uk/BiogasChina.php> [accessed 21 July 2014].

Karki, A.B., Gautam, K.M. and Kandel, G. (1996) *Biogas Technology: A training manual for extension, Food and Agricultural Organisation*, Consolidated Management Services Nepal, Kathmandu, Nepal. <www.fao.org/docrep/008/ae897e/ae897e00.htm> [accessed 23 July 2014].

Karki, A.B. and Ghimire, P.K. (1998) *Sustainable Approach on Quality Control of Biogas Plants*, Consolidated Management Service Nepal (P) Ltd. for AEPC. <www.snvworld.org/download/publications/sustainable_approach_on_quality_control_nepal_1998.pdf > [accessed 21 July 2014].

Karki, A.B., Shakya, I., Dawadi, K.D. and Sharma, I. (Eds) (2007) *Biogas Sector in Nepal: Highlighting historical heights and present status*, Nepal Biogas Promotion Group (NBPG), Kathmandu, Nepal.

Khandelwal, K.C. (2008) *Country Report on Financing of Domestic Biogas Plants in India,* SNV, Jaipur India. <www.snvworld.org/en/Documents/India_international_workshop_on_financing_of_domestic_biogas_plants_2008.pdf > [accessed 21 July 2014].

Lam, J. (1996) *Enforcement of Quality Standards upon Biogas Plants*, Kathmandu, Nepal: BSP (Biogas Support Programme), SNV. <www.snvworld.org/en/download/publications/enforcement_quality_standards_nepal_1996.pdf > [accessed 21 July 2014].

Leermakers, M. (1992) *Extension of Biogas in Nepal Theory and Practice,* : SNV Nepal, Kathmandu, Nepal <www.snvworld.org/download/publications/

extension_of_biogas_in_nepal_theory_and_practice_1992_0.pdf>[accessed 21 July 2014].

Moulik, T.K. (1985) *The Biogas Programme in India and China: A comparative analysis of experiences*, Ahmedabad: Indian Institute of Management.

van Nes, W. (2007) 'Commercialisation and business development in the framework of the Asia Biogas Programme', in Seminar on Policy options for expansion of community-driven energy service provision, Beijing, China.

van Nes, W., Lam, J., ter Heegde, F. and Marree, F. (2009) Building viable domestic biogas programmes: success factors in sector development, SNV Netherlands Development Organisation. <www.snvworld.org/files/ publications/snv_building_viable_domestic_biogas_programmes_2009. pdf> [accessed 21 July 2014].

NIIR, B. (2004) *Handbook on Biogas and its Applications*, Delhi: National Institute of Industrial Research < www.niir.org/books/book/handbook-on-bio-gas-its-applications/isbn-8186623825/zb,,72,a,5,0,a/index.html> [accessed 21 July 2014].

Peters-Stanley, M. and Yin, D (2013) *Manoeuvring the Mosaic - State of the voluntary carbon markets 2013*, Ecosystem Marketplace (Forest Trends and Bloomberg New Energy Finance) Washington DC. <www.forest-trends.org/ documents/index.php?pubID=3898 > [accessed 21 July 2014].

Shah, O.P. (2006) *Sustainable Waste Processing in Mumbai: Using the Nisargruna Technology*, MSc Thesis. Borås, Sweden: Hogskolan i Borås. <www. kingdombio.co.uk/ShahThesis.pdf > [accessed 21 July 2014].

SNV (2006) *Biogas Project Evaluation Report*, Hanoi, Vietnam, SNV Vietnam. <www.snvworld.org/en/download/publications/vietnam_biogas_project_ evaluation_report_2006.pdf> [accessed 23 July 2014].

Tumwesige, V., Fulford, D. and Davidson, G.C. (2014) 'Biogas appliances in Sub-Sahara Africa', *Biomass and Bioenergy* (In Press) < http://dx.doi. org/10.1016/j.biombioe.2014.02.017> [accessed: 21 July 2014].

Voorhes, M., Humphreys, J. and Solomon, A. (2012) *Report on Sustainable and Responsible Investing Trends in the United States*, US SIF, The Forum for Sustainable and Responsible Investment, Washington DC. <http://ussif. membershipsoftware.org/files/Publications/12_Trends_Exec_Summary. pdf> [accessed 21 July 2014].

Zhang, M. (2008) *Financing of Domestic Biogas Plants in China*, SNV, Bangkok, Thailand. <www.snvworld.org/en/Documents/China_international_work-shop_on_financing_of_domestic_biogas_plants_2008.pdf> [accessed 21 July 2014].

CHAPTER 11
Starting a biogas programme

Abstract

Starting a new biogas extension programme in an area requires information that should be gained from an assessment survey. People in the area need to know about the benefits of biogas, so publicity is required. A limited pilot programme allows staff to be trained and extension methods to be tested and established. As the programme moves forward standards should be defined that are used to encourage quality control. Follow-up surveys quickly identify areas in which the programme management can be improved. Staff need to be flexible so that improvements can be made and new opportunities recognized.

Keywords: biogas, anaerobic digestion, starting an extension programme

Preparations for a biogas programme

When a programme is started in a country or region in which biogas, or other new technology, has not been used previously, all of the factors involved in managing the programme need to be put in place as it is being established. This includes the provision of an effective technology by people who understand its installation and use, and who know how to train others in its extension (Kudaravalli, 2014).

A biogas programme needs to be appropriate for the area in which it is to be run. The technology should use local resources and not rely on components that are imported from elsewhere. The skills to build and run biogas plant designs should be those available locally, or for which local people can easily be trained. An over-reliance on external expertize has been a major cause of failure for many development projects in the past.

Assessment surveys

When an organisation involved in biogas extension programmes, such as the Netherlands Development Orgnisation (SNV) and SKG Sangha, plans a biogas project in a new country or region, the first task is a feasibility study (see Appendix 5). Such a study provides key baseline information that allows a plan to be made for the establishment of the new programme (Ghimire, 2011). A request for a biogas extension programme has often come from national or local government, an aid group, or an entrepreneur who sees the potential for the technology in the area. One or more experts from an existing

http://dx.doi.org/10.3362/9781780448497.011

biogas programme will visit the area and evaluate the potential for a biogas programme there (Teune *et al.*, 2010). The outside expert(s) will work with key local people to obtain answers to basic questions about the area. Potential stakeholders should be identified and consulted. One key stakeholder is the group that is funding such a survey; whether it is a government department, aid organisation, or local entrepreneur. Stakeholders have particular expectations from a biogas extension programme and the survey needs to confirm the extent to which these expectations can be realized.

The first dimension to consider is that of socio-technology: what people do and how they do it. For domestic biogas, a key aspect of this dimension is the way that women cook. This includes the fuel that is used, the type of cooking stoves used, and the cultural attitudes and practices that surround cooking. The reason that biogas became so popular in India and Nepal was that the cooking gas was made from cow dung (gobar) because cattle have a particular place in local society.

If animal dung is to be the main feed material, it is important to assess the proportion of local people who own animals and the average number of animals per household. The type of animal ownership is also an issue. If cattle are mainly owned by nomadic pastoralists, rather than settled farmers, establishing a biogas programme may be very difficult. If other biomass residues are to be the feed material, there is a need to assess the quantities of residues that are available in total, but also the average amounts that are available to each family.

The local climate is also a factor. The low-cost biogas systems used in the large biogas programmes are not designed with heating because the climate is normally at a high enough temperature. Work has been done in Nepal to develop biogas systems for the lower temperatures found at higher altitudes, but this needs further research and development. Other local aspects are ground conditions and the availability of water. Low-cost biogas systems are designed to be built underground, but if the ground is too wet, or too hard, this may not be possible. Water is required to mix the dung, so a biogas programme may place an additional pressure on a possible scarce resource.

The potential benefits of a biogas programme to an area should be assessed; they are part of the motivation for people to accept the use of biogas plants. The health of people is one issue and should be part of the survey. If people are cooking over wood fires, there may be issues with smoke related diseases. The use of people's time, especially that of women, is another issue. If women have to spend a long time each day collecting firewood, the provision of a biogas plant will allow them to spend much less time cooking. The opportunities for using the extra time available, such as for money earning activities, can be investigated. Other benefits may come from association with the project. For example, the biogas-extension workers may also be able to install piping for water while they are building a biogas plant.

The amount of fossil carbon saved is another calculation for which data needs to be sought. If carbon finance is a potential source of income for the project, then such information is essential for making the case for finance.

The survey work itself will need to be done by a team of people, most of whom will be recruited locally. The finance for the survey will come from the organization that has instigated the idea for a biogas extension programme in the area, such as a government department, aid organisation or local entrepreneur. It is important to ensure that adequate finance is available for this survey because accurate information at the initial stages will enable effective plans to be made. Good planning will save expensive mistakes later in the project.

Advertising and motivation

When biogas technology is introduced into a new area, it is a new concept for people, so they need to be informed about its benefits. This can be done through local media, such as radio and newspapers, depending on how people receive information. Different groups have identified novel ways to inform people of the technology. Biotech Ltd fitted a working model of a biogas plant on the back of a small truck, which could be taken to schools and local agricultural fairs, and launched it on local TV. Villages in Vietnam had been fitted with public speakers over which propaganda could be broadcast. These systems are presently used for local education and delivering information and provided a good medium with which to publicize biogas technology (Fulford *et al.*, 2012). The increasing use of the mobile phone in many parts of the world suggests that mobile phone adverts could also be used.

Extension workers, who will explain the benefits of biogas technology, can target community leaders. The extension workers are encouraged to set up local meetings in which the message can be spread to a wider number of people. The meetings can include entertainment and food to encourage people to attend. Young people are often more ready to learn about new ideas, so they can be recruited to tell other people about the meetings. Visits to schools can be a way to inform such young people and enthuse and encourage them to tell their parents, friends and neighbours.

The most effective advertising is by word-of-mouth and the most enthusiastic advertizers are people who have a new biogas plant. Once the programme has started, people can be encouraged to visit new plants to see them running. Even before any plants are built in a particular area, a tour can be organized for a group of people to visit plants in a area with an established programme.

Pilot programme

Once detailed plans can be made for a biogas extension programme, staff must be hired and trained for the work. Initial training can be done within a successful biogas extension project in another area. Final training needs to be done 'on the job' while building biogas plants in the programme area. During this time, staff will take longer to do tasks, until they become routine, and will make mistakes which will need to be rectified.

The answer is a pilot programme (Fulford, 1988), which allows the programme to be developed slowly. This phase of the programme will be more expensive and difficult to run than the full programme and plans must be made to allow for this. Provision must be made for this learning phase. Not all the staff that join the programme will stay in it, so more staff need to be trained than will be needed to run the programme. Some will find a new programme too difficult to handle, others will accept the training and then leave to set up their own work, either in competition, or in another field.

Technical mistakes will be made, so a guarantee fund should be set up to enable mistakes to be rectified. Repairs to failed plants should be made quickly because a delay will be seen as a failure of the technology. However, customers should be expected to pay for the technology they receive, as long as it works and provides the service that they expect. Installing new technology for reduced or zero cost means that customers do not fully accept what they are given.

There are people in any society who are keen to try out a new technology, so if guarantees are provided, these people will want to become involved. There is an argument for providing a free demonstration unit, in which people can test out the technology for themselves. A better approach is to provide the first system under a deferred payment agreement, so the customer can test the technology and only pay when they are happy that it does the job that the installer promised it would. If the first customers are expected to pay the standard amount for what they receive, the rest of the society will develop the right expectations and be happy to join in, once they have developed confidence in the technology.

The pilot programme should have a defined scope, with a limited number of units installed in a limited time. If a large number are installed and problems are found in the design, a greater number of repairs must be made. For example, in Nepal, the pilot programme accepted by DCS for the floating drum design, installed 95 plants (out of a target of 100). The floating drums rusted more quickly than expected, so extra money had to be requested from the funding agency to replace most of these drums with new ones that were properly painted. A smaller number of plants would have allowed a reduced cost for their replacement. An important part of the pilot programme is a thorough follow-up and analysis of the plants installed to check on quality and effectiveness of the technology. If problems are identified at this stage, they can be corrected before the main programme is established.

Main programme

As staff gain experience and the technology becomes more reliable, the pilot scheme can merge into an on-going extension programme. However, there is a temptation to expand the programme quickly and this should be resisted.

The Chinese programme expended very rapidly in the 1970s and there were a large number of failed plants (over 57% by 1978 (Karki *et al.*, 1996)). New staff need thorough training in all aspects of the work. Again, more staff need to be trained than the programme actually requires. This allows the programme to expand more quickly than expected, as well as allowing staff to drop out.

The need for thorough follow-up and analysis of the plants installed under the pilot programme shows that research and development is an important aspect of a new programme. People with experience of technical research and development are able to recognize problems as they occur and find ways to solve them. They are also able to train the extension staff in the technical aspects of their work. Effective follow-up is essential to ensure the success of the programme. If problems are identified quickly and put right, people gain confidence in the technology and are willing to trust the programme. This increases the demand for biogas plants.

An essential part of the pilot programme is the establishment of standards for plant design and provision of suitable gas appliances and equipment (Chen *et al.*, 2012). Once these standards have been tested, they need to be written down and supplied to all of the construction staff (Lam, 1996). The standards for the plant design can be provided as a list of key dimensions to which masons should adhere. A quality-control evaluation by supervisors can then be provided as a check list that each of these measurements is correct (Karki *et al.*, 1996).

A range of pipe valves and fittings will have been tested during the pilot programme and the most cost-effective ones identified. The on-going programme will ensure that these particular supplies are made available and used in the construction of plants. The same is true of gas appliances, such as biogas burners and lights. A quality control check will ensure that the correct items have been used. Some installers may try to cut corners and use low-quality gas equipment, so it is important to ensure the right equipment is being used.

Follow-up surveys

The use of regular follow-up surveys (see Appendix 6) is closely linked to good quality-control and the success of a biogas programme. The main reason for failures in some biogas programmes in India and elsewhere is that plants were installed and the users were then ignored. Some installers did not finish the installation work, so customers were lacking a gas pipeline, or the plant leaked gas because the dome was not properly sealed. Customers were not given adequate training in the use of their plants, so did not know how much feed material to use, or how to prepare it. Biogas plants may have minor faults which are easy to repair, such as the accumulation of water in the gas pipeline. If customers do not know the cause of these faults and how to overcome them, they may stop using the plant and assume that it has failed.

Supervisors can do regular follow-up surveys, but it is important that an independent team also make occasional follow-up visits to a sample of the plants being built. A group of supervisors in a particular area may accept lower-quality standards than is required (Fulford *et al.*, 1991), so checks need to be made. An independent group may also be able to identify particular issues in the standard design that need to be corrected. All staff need to be encouraged to make suggestions as to how designs or procedures could be made more cost-effective. Independent follow-up visits can check these suggestions and consider whether some of them could be adopted.

These independent follow-up surveys also need to ask questions of the customers, to ensure the programme meets their needs. Changes in design or procedures may make a biogas system easier to use by the customers, but these may not be recognized by the supervisors or masons. The expectations of customers also need to be managed.

Each SNV biogas programme has attempted to do follow-up surveys each year, both by internal staff and by external consultants. Examples of such surveys are available for programmes in places such as Nepal (NESS, 2011), Bangladesh (iDE, 2011) and Vietnam (Dung *et al.*, 2009). Such surveys allow these programmes to identify any weaknesses, so they can be corrected to ensure that the programmes continue to be successful.

Flexibility

As the programme moves beyond the start-up phase and the approach and technology become more routine, it is important that the staff are willing to make changes, if they are required. It is very easy for a biogas extension programme to become too rigid, so that people do not recognize new opportunities (Fulford *et al.*, 2012). Managers should encourage all staff to make suggestions for improvements and be willing to take these ideas seriously. If an attitude of research and development is part of the culture of a programme, the technology and approach can be adapted as new opportunities become apparent.

References

Chen, L., Zhao, L., Ren, C. and Wang, F. (2012) 'The progress and prospects of rural biogas production in China', Renewable Energy in China 51: 58–63 <http://dx.doi.org/10.1016/j.enpol.2012.05.052> [accessed 21 July 2014].

Dung, T.V., Hung, H.V. and Hoa, H.T.L. (2009) Biogas User Survey, Vietnam 2007–2008, Independent Consultant Group for SNV, Hanoi, Vietnam. <www.snvworld.org/en/download/publications/biogas_user_survey_livestock_2009_vietnam.pdf> [accessed 21 July 2014].

Fulford, D. (1988) Running a Biogas Programme: A handbook, London: Practical Action Publications (Intermediate Technology Publications). <http://developmentbookshop.com/running-a-biogas-programme-pb> [accessed 22 July 2014].

Fulford, D., Devkota, G.P. and Afful, K. (2012) Evaluation of Capacity Building in Nepal and Asia Biogas Programme, Hanoi, Vietnam: Kingdom Bioenergy Ltd for SNV. <www.kingdombio.com/Final Report - whole.pdf> [accessed 21 July 2014].

Fulford, D., Poudal, T.R. and Roque, J. (1991) Evaluation of On-going Project: Financing and construction of biogas plants, United Nations Capital Development Fund, New York.

Ghimire, P.C. (2011) Feasibility Study of National Domestic Biogas Programme in Sri Lanka, People in Need, SNV for PIN, Colombo, Sri Lanka. <www.snvworld.org/en/download/publications/biogas_feasibility_study_sri_lanka_2011.pdf> [accessed 21 July 2014].

iDE (2011) Annual Biogas User Survey, 2010, IDCOL and NDBMP, Dacca Bangladesh. <www.snvworld.org/en/download/publications/biogas_user_survey_2010_bangladesh_2011.pdf> [accessed 21 July 2014].

Karki, A.B., Gautam, K.M. and Kandel, G. (1996) Biogas Technology: A training manual for extension, Food and Agricultural Organisation, Consolidated Management Services, Kathmandu Nepal. <www.fao.org/docrep/008/ae897e/ae897e00.htm> [accessed 23 July 2014].

Kudaravalli, K.K. (2014) Extension Manual for Biogas Technology, Kolar, South India: SKG Sangha. <http://foundationskgsangha.org/Extension Manual.pdf> [accessed 21 July 2014].

Lam, J. (1996) Enforcement of Quality Standards upon Biogas Plants, Kath mandu, Nepal: BSP (Biogas Support Programme), SNV. <www.snvworld.org/en/download/publications/enforcement_quality_standards_nepal_1996.pdf> [accessed 21 July 2014].

NESS, N.E.& S.S. (2011) Biogas User Survey, Nepal 2009–2010 (CDM Activity 1), AEPC.<www.snvworld.org/en/download/publications/biogas_user_survey_2009-2010_cdm_activity_1_nepal.pdf> [accessed 21 July 2014].

Teune, B., Orprecio, J., Dalusung, A. and Yeneza, G. (2010) Feasibility Study of a National Biogas Programme on Domestic Biogas in the Philippines, SNV and Winrock International Little Rock, Arkansas, USA. <www.snvworld.org/en/download/publications/feasibility_study_of_a_national_domestic_biogas_programme_the_philippines_2010.pdf> [accessed 21 July 2014].

CHAPTER 12
Philosophy of biogas extension

Abstract

The concept of the triple bottom line encourages the application of auditing to the environmental and social aspects of a project, as well as to the financial aspects. Biogas extension programmes have a high potential to demonstrate that they are sustainable in all three areas: environmental, social and financial, if they are properly planned and managed. The use of biogas to process food waste allows it to change the lives of the poorest of the poor.

Keywords: biogas, anaerobic digestion, philosophy, triple bottom line

Triple bottom line

An effective extension programme needs to be sustainable in terms of least three aspects, otherwise it cannot continue (Foran *et al.*, 2005; Hacking and Guthrie, 2008). Firstly, it needs to be financially sustainable at several levels. People need to be able to afford the technology and gain financial benefits from its use. The programme itself, whether run as a non-profit NGO or as a commercial operation (or operations) must earn sufficient income to pay the people doing the work.

Linked with the financial benefits that people feel they are gaining from using the technology, is a sense of social benefit. People need to feel that the extension programme is socially sustainable. The lifestyles of people involved in the programme, both as customers and workers, should be improved by their involvement. The opposite of this approach has been used by some development programmes that have exploited people. Too many groups have used the time, effort and other resources of one local group of people to benefit others that do not belong to that local community.

The third area in which a project is seen as sustainable is the environment. A project should not exploit local environmental resources in a way that destroys them. Again, there have been too many projects that have caused degradation to local resources, such as an overuse of wood-fuel leading to deforestation. Development projects can lead to an increase in the demand for energy, which can result in a greater use of fossil fuels and an increase of the release of fossil carbon into the atmosphere.

Environmental sustainability of biogas technology

Biogas is seen as a sustainable technology. When biogas is burnt, it does generate carbon dioxide, but the carbon originally came from growing plants.

http://dx.doi.org/10.3362/9781780448497.012

The animal dung that is used as a feed material is the result of animals digesting the plant material that they have eaten. Most food is made from plants that have grown by absorbing carbon dioxide from the atmosphere. A small amount of fossil carbon may have been involved in transporting and processing some foods, so the biogas process is not 100% sustainable, but it is very close. If a biogas plant replaces the use of dung heaps and food-composting systems, methane is captured and burnt instead of being released into the atmosphere. Since methane is seen as being between 17 and 30 times worse (depending on assumptions) as a greenhouse gas, than the carbon dioxide released when it is burnt, the use of biogas technology can be seen as reducing the effects of climate change.

The production of a high-quality compost can also be included in the carbon offset equation because compost replaces inorganic fertilizer that requires fossil carbon for its manufacture. Some of the carbon in plant material added to the soil as compost degrades very slowly, so it has a fairly long lifetime. The use of compost made from biogas plant effluent can therefore be seen as trapping carbon and retaining it over long periods of time, which is another positive environmental benefit. The use of high-quality compost can also encourage better crop growth, which traps carbon dioxide from the atmosphere, particularly if the compost is used to encourage the growth of trees.

A full life cycle analysis (LCA) of a biogas programme will take into account that biogas plants are made from concrete or masonry. Such analyses have been done for biogas systems in Europe (Berglund and Börjesson, 2006; Börjesson and Berglund, 2007, 2006), but analyses for domestic plants in Asia are very limited. The manufacture of cement from limestone or chalk (calcium carbonate) and its use in concrete or mortar does release fossil carbon dioxide. The manufacture of bricks requires that they be fired, using a source of energy. If coal is used to provide the heat, an amount of fossil carbon must be added to the equation. If biomass is used to fire bricks, the process could be considered to be more sustainable, but only if the biomass is taken from an energy plantation, rather than forest areas that are not being replanted. A masonry biogas unit should last for many years, so the fossil carbon used in building a plant can be amortized over the lifetime of the plant. The yearly carbon cost will be offset by much greater yearly carbon benefits.

The main benefits of the use of biogas come from its use as a replacement for other fuels. Where biogas is used as a fuel in engines or as a cooking fuel in place of LPG, it directly replaces fossil fuels. Where biogas replaces wood as a cooking fuel, it reduces deforestation. The environmental benefits of biogas extension programmes have been recognized by groups such as World Wildlife Fund (WWF). In one area of Nepal, WWF has supported the installation of 7,500 biogas units and they claim they have saved 33,000 tonnes of firewood a year, or 617 acres of forest (WWF Nepal, 2012).

Social sustainability

The benefits of biogas technology for people's lifestyles are listed in Chapter 3, including the benefits to people's health (Mahoney and Potter,

2004). Where biogas replaces wood as a cooking fuel, it reduces the amount of smoke in people's kitchens. As a consequence, it reduces the incidence of eye, nose, throat, and lung disease in woman and children. Women feel they are much cleaner because they do not smell of wood smoke. Their pots and pans can be cleaned more easily because there is no soot from biogas flames. While animal dung does need to be collected to feed a biogas plant, this is much easier than having to travel to the forest to collect firewood. People need to go to their animal sheds each day to check and feed the animals, so the collection of dung is a small extra task to do. Firewood is heavy and often has to be carried for long distances, so a biogas plants reduces back pain and muscle strain.

Visits to people with biogas plants often reveal unexpected extra benefits. Women claim they can feed their children with a hot meal before school, which enables them to study more effectively. The children are also less tired when they go to school because they do not need to collect firewood. Using latrines connected to biogas plants has improved sanitation; previously people had to use the fields. Women often have to go out at dawn and dusk when snakes and other dangerous creatures are active, so a latrine attached to a biogas plant improves their safety.

In the past, the use of biogas for domestic purposes in rural areas has been criticised as not helping the poorest of the poor. People have to own land and animals in order to use a biogas plant. The development of biogas plants to process food wastes can overcome this issue, as the poorest sections of a community are often those who recycle wastes.

In urban areas, where many of the poorest people live, waste is often collected as mixed materials and put on dumps. People who try to make a living by separating the waste materials into plastics, glass, metals, paper and fabrics to sell them need to clean these valuable materials and remove rotting materials, especially food wastes. The rotting materials generate obnoxious smells, are toxic and cause disease. If food waste can be used to generate biogas and compost, it gains an economic value. If institutions and local authorities are willing to set up a food-waste processing system, the poor people doing the work of recycling can be employed to collect food waste to generate energy and good-quality compost. The other waste materials are less contaminated, so are easier to recycle.

The use of biogas to recycle food waste to energy and compost has a high potential to improve the social sustainability of the poorest of the poor.

Financial sustainability

A biogas extension project has to be financially sustainable at two levels. The customers need to have confidence that their investment in a biogas plant can be recovered within a few years through savings or through extra income. The biogas extension programme must earn enough to pay its staff to keep running.

Most customers need to borrow money to pay the capital cost, from a bank or an MFI (Micro-finance Institute), so the savings or earning must be sufficient to cover the loan repayments. Chapter 3 gives details of the studies made on the economics of biogas systems over the years. When the large programmes were started in Asia in the 1970s, a simple financial analysis of domestic biogas systems gave a marginal answer. If a family purchased a biogas plant, they saved just about enough money to cover the costs of a loan. A growing scarcity of firewood and an increase in its market cost in many places plus the increasing cost of inorganic fertilizer has changed the balance to make it much more positive. If biogas is used to replace LPG as a cooking fuel, the financial savings appear to be much greater, making the benefit:cost ratio much higher.

If an institution sets up a biogas plant to process food waste, the financial benefits are usually positive. The calculation depends on the capital cost of the plant and the cost of the LPG that it replaces. If the biogas is used to generate electricity in place of diesel, the benefits are found to be about the same when calculated in an Indian context. The benefit:cost ratio can be further increased by linking the biogas to latrines, so biogas can be generated from human sewage as well as food wastes. In many places, though, cultural pressures may prevent uptake of latrines linked to biogas production.

A biogas extension programme needs to be financially sustainable in order to keep running. Many biogas programmes combine both non-profit NGO management components with commercial organisations. There are arguments that the sales and construction of biogas plants at the local level should be run on a commercial basis (van Nes *et al.*, 2009). An NGO approach to biogas extension has also been very successful, as SKG Sangha have demonstrated in South India (Ashden, 2007). The biogas extension programme in Bangladesh involved tension between NGO and commercial thinking (Fulford *et al.*, 2012). IDCOL established the programme with the idea that commercial companies would take contracts to build biogas plants. Many of the groups that were already building plants were NGOs, such as Grameen Shakti (Ashden, 2008). They found it difficult to think commercially, so consistently failed to meet numerical and financial targets. The NGOs already had sources of external funding, so did not understand the need to meet these targets.

However an extension programme is run, it is important to have a business plan in which sources of income and expenditure are clearly identified and projections made for several years into the future. If the plan indicates that the project is not financially sustainable, efforts can be made to increase income (e.g. by looking for carbon offset contracts) or by reducing costs. The time taken to develop a good business plan, even if this takes several months, will save a large amount of time (many years) in the future operation of the programme, as decisions can be made on the basis of a good understanding of the consequences.

A financial bottom line also includes consideration of resources other than finance that need to be used in running the programme. The people employed in a programme and the skills they offer it also need to be carefully planned. The business plan should include the training of staff and also the ability of staff to advance within the organisation if they demonstrate a commitment to the work and an ability to grow their skill level. A weakness of several programmes has been the loss of trained masons when they realise they can earn more income elsewhere with the skills they have gained from being trained. This was not seen as an issue in Vietnam (Fulford *et al.*, 2012) because the masons run the construction companies. Few of the mason-run construction groups are officially registered as businesses, so they operate in the informal economy. The masons are very loyal to the biogas extension programme because they feel they have control over their work and see the opportunities to develop their work for the future.

Wider benefits of biogas programmes

A deeper consideration of the benefits of biogas technology suggests a quadruple bottom line (Sood and Tulchin, 2013). The fourth area in which a programme could be considered sustainable is spiritual (Welty, 2014). The ability of a biogas plant to convert obnoxious materials, such as animal dung and rotting food wastes, into valuable products, such as energy and compost, can be seen in many religions as a parable of transformation. The technology has been shown to be very attractive to ashrams such as ViveKananda Kendra in Kanya Kumari (Ashden, 2006) in south India and Shri Kshethra in Dharmasthala (Ashden, 2012) in the Western Ghats of India. The technology reflects philosophies that are deeply concerned for the environment and encourage people to care for it.

The concept of transforming the unclean into something positive and beneficial is traditionally represented by the lotus flower. It grows in obnoxious smelling swamps, but is both beautiful and sweet smelling. This image could be used as a symbol for biogas technology.

In Bangladesh, Grameen Shakti (Ashden, 2008) added the extension of biogas plants to its very successful programme of installing solar systems in homes. The aim was to give local people a greater range of ways to improve their lives, by offering clean cooking fuel as well as a source of solar light. The list of customers of Biogas Technology Africa Ltd (Idan, 2006) includes schools and universities established by local and international churches.

Considerations from the triple bottom line

Biogas technology extension programmes came out of a concern that development based on the exploitation of fossil resources was limited and an approach based on the use of renewable resources was needed. The ability of a biogas system to transform obnoxious wastes, such as animal (and human) dung and rotting food residues, into good-quality fuel gas and compost makes it an ideal example of this alternative approach. Over four decades, biogas extension programmes have been able to demonstrate that they are financially, environmentally, and socially sustainable. Biogas extension programmes offer a clear positive balance on the triple bottom line, even on the quadruple bottom line, because they are also spiritually sustainable.

References

Ashden (2006) 'VK Nardep, India: Adding value to the residue from biogas plants', [website – case study] Ashden <www.ashden.org/winners/vknardep> [accessed 21 July 2014].

Ashden (2007) 'SKG Sangha, India, Biogas for cooking plus fertiliser from slurry', [website – case study] Ashden <www.ashden.org/winners/skgsangha> [accessed 21 July 2014].

Ashden (2008) 'Grameen Shakti, Bangladesh: Rapidly growing solar installer also provides clean cooking', [website – case study] Ashden <www.ashden.org/winners/grameen08> [accessed 21 July 2014].

Ashden (2012) 'SKDRDP, India: Enabling the poor to make informed energy choices', [website – case study] Ashden <www.ashden.org/winners/skdrdp12> [accessed 21 July 2014].

Berglund, M. and Börjesson, P. (2006) 'Assessment of energy performance in the life-cycle of biogas production', Biomass and Bioenergy 30: 254–266 <http://dx.doi.org/10.1016/j.biombioe.2005.11.011> [accessed 21 July 2014].

Börjesson, P. and Berglund, M. (2006) 'Environmental systems analysis of biogas systems, Part I: Fuel-cycle emissions', Biomass and Bioenergy 30: 469–485 <http://dx.doi.org/10.1016/j.biombioe.2005.11.014> [accessed 21 July 2014].

Börjesson, P. and Berglund, M. (2007) 'Environmental systems analysis of biogas systems, Part II: The environmental impact of replacing various reference systems', Biomass and Bioenergy 31: 326–344 <http://dx.doi.org/10.1016/j.biombioe.2007.01.004> [accessed 21 July 2014].

Foran, B., Lenzen, M., Dey, C. and Bilek, M. (2005) 'Integrating sustainable chain management with triple bottom line accounting', Ecological Economics 52: 143–157 <http://dx.doi.org/10.1016/j.ecolecon.2004.06.024> [accessed 21 July 2014].

Fulford, D., Devkota, G.P. and Afful, K. (2012) Evaluation of Capacity Building in Nepal and Asia Biogas Programme, Hanoi, Vietnam: Kingdom Bioenergy Ltd for SNV. <www.kingdombio.com/Final Report - whole.pdf> [accessed 21 July 2014].

Hacking, T. and Guthrie, P. (2008) 'A framework for clarifying the meaning of triple bottom-line, integrated, and sustainability assessment', Environmental Impact Assessment Review 28: 73–89 <http://dx.doi.org/10.1016/j.eiar.2007.03.002> [accessed 21 July 2014].

Idan, J.A. (2006) 'Sustainable Sewage and Biodegradable Solid Organic Waste-to-Energy Technology', [website - presentation] <www.biogasonline.com/downloads/suspbeAfrica.pdf> [accessed 21 July 2014].

Mahoney, M. and Potter, J.-L. (2004) 'Integrating health impact assessment into the triple bottom line concept', Environmental Impact Assessment Review 24: 151–160 <http://dx.doi.org/10.1016/j.eiar.2003.10.005> [accessed 21 July 2014].

van Nes, W., Lam, J., ter Heegde, F. and Marree, F. (2009) Building viable domestic biogas programmes: success factors in sector development, SNV Netherlands Development Organisation, The Hague, Netherlands. <www.snvworld.org/en/download/publications/snv_building_viable_domestic_biogas_program mes_2009.pdf> [accessed 21 July 2014].

Sood, S. and Tulchin, D. (2013) Quadruple Bottom Line, Social Enterprise Associates Tip Sheet, Washington DC #13 <www.socialenterprise.net/assets/files/TipSheet13QBL.pdf> [accessed 21 July 2014].

Welty, C. (2014) 'Chapter 22 - The Regenerative Community Régénérer: A Haitian Model and Process Toward a Sustainable, Self-Renewing Economy', in W.W. Clark, (ed.), Global Sustainable Communities Handbook, pp. 527–538, Boston: Butterworth-Heinemann. <www.sciencedirect.com/science/article/pii/B9780123979148000229> [accessed 21 July 2014].

WWF Nepal (2012) 'Biogas comes to Madhuban: a story of change', [website - presentation] WWF, Kathmandu, Nepal. <http://awsassets.panda.org/downloads/biogas_booklet.pdf> [accessed 21 July 2014].

APPENDIX 1
Chemistry of simple digestion

Abstract

A basic analysis of the process by which food materials are converted to biogas provides an understanding of the energy released by this process. Calculations are provided for the conversion of two basic food materials: sugar and starch. A simple first-order rate-model is also presented, which allows the gas production from a biogas plant to be estimated, based on the plant working volume and the daily feed volume.

Keywords: biogas, anaerobic digestion, chemistry, thermodynamics

Process of conversion of food materials to biogas

Food materials are composed of carbohydrates, fats, and proteins. As explained in Chapter 4, in anaerobic digestion, the insoluble components are turned into simpler soluble chemicals by hydrolysis. These soluble chemicals are then processed to long-chain fatty acids and then simpler fatty acids by acidogenesis and acetogenesis. Methanogenesis then takes the short-chain fatty acids and turns them into methane and carbon dioxide.

The details of the process are complex; a wide range of different microorganisms (bacteria and archaea) are involved at each stage of the process. However, it is possible to start with simple food materials, such as sugar and starch, and work out a mass and energy balance for the overall process. The details of this analysis were originally determined by Arthur Buswell (Buswell and Neave, 1930), who showed that the composition of biogas is dependant on the amounts of different food materials in the feed material (such as carbohydrate, protein and fats).

An important parameter in determining the amount of biogas that a feedstock can produce is the chemical oxygen demand (COD), the amount of oxygen that it requires to break it down. For simple food materials, where the chemical formula is known, it is possible to determine a theoretical value for the COD. By comparing the actual biogas production to the theoretical, it is possible to determine an overall efficiency for the anaerobic digestion process.

The mass and energy balances and the COD determination for sugar and starch are determined in the sections below.

http://dx.doi.org/10.3362/9781780448497.013

Calculations for sugar $C_{12}H_{22}O_{11}$

One of the simplest foodstuffs is sugar, which is most common in the form of a disaccharide, for which the basic equation is defined: giving a molecular weight of 342. If sugar is burnt in a stochiometric mix with air, it forms carbon dioxide and water:

$$C_{12}H_{22}O_{11}+12O_2+45.14N_2 \rightarrow 12CO_2+11H_2O+45.14N_2$$

The mass balance for this equation (ignoring the nitrogen) is given below.

Component	$C_{12}H_{22}O_{11}$	$12O_2$	→	$12CO_2$	$11H_2O$
Molecular weight	342	32		44	18
Mass (kg)	342	+384	=	528	+198

The nitrogen does not take part in the reaction, but is included because we are assuming the sugar is burnt in air.

The mass balance shows that a mass of 342 kg of sugar requires 384 kg of oxygen to be consumed, or 1 kg of sugar requires 1.123 (=384/342) kg of oxygen. The theoretical COD of sugar is therefore 1.123 kg/kg, or 1.123 g/g.

To define an energy balance, the gross[a] enthalpy of combustion of sugar is required, which is: −5,644,170 kJ/kg mole (NIST, 2011). When sugar is burnt in air, it releases 16.503 MJ/kg.

Working from this figure, it is possible to calculate the energy of formation of sugar. The energy of formation of carbon dioxide is −393,510 kJ/kg mole and that of liquid water[b] is -285,830 kJ/kg mole (NIST, 2011). The enthalpy of formation of elements in their natural form, such as oxygen and nitrogen as diatomic gases, is zero.

Component	$C_{12}H_{22}O_{11}$	$+12O_2$	→	$+12CO_2$	$+11H_2O$	Total
Energy (kJ/kg mole)	−2,222,080	0		−393,510	−285,830	−5,644,170
Total	−2,222,080	0	=	−4,722,120	−3,144,130	−5,644,170
Energy (kJ/kg)	−6,497	0		−8,943	−15,879	−16,503

If sugar is put in an anaerobic digester, it is broken down into carbon dioxide and water, with the addition of a water molecule by hydrolysis:

$$C_{12}H_{22}O_{11}+H_2O \rightarrow 6CO_2+6CH_4$$

A similar mass and energy balance can be defined (using the enthalpy of formation of sugar derived from the previous table – see (NIST, 2011)).

Component	$C_{12}H_{22}O_{11}$	$+H_2O$	→	$6CO_2$	$+6CH_4$	Total
Molecular weight	342	18		44	16	
Mass (kg)	342	+18	=	264	+96	
Energy (kJ/kg mole)	−2,222,080	−285,830		−393,510	−74,870	
Total	−2,222,080	−285,830	=	−2,361,060	−449,220	−302,370
Energy (kJ/kg)	−6,497	−15,879		−8,943	−4679	−884.1

This calculation shows that the generation of biogas is exothermic (884.1 kJ/kg sugar), but the energy generated is very low compared to that generated when sugar is fully converted to carbon dioxide and water (5.36%=0.8841/16.503).

1 kg of sugar gives 1.053 kg of biogas (50:50 methane and carbon dioxide), which contains 0.281 kg methane. Burning this methane would generate 15.619 MJ. Adding this figure to that generated from the conversion of sugar to biogas gives 16.503 (=15.619+0.884) MJ, so the energy calculations balance.

The volume occupied by the biogas produced by 1 kg of sugar is 0.839 m³ (at NTP^c), of which 0.4195 m³ is methane. When biogas is burned, it requires oxygen from the air. The COD of a 50:50 mix of biogas is 1.067 g/g (=32/30) as shown in the mass balance below:

$$0.5CO_2 + 0.5CH_4 + O_2 + 3.762N_2 \rightarrow CO_2 + H_2O + 3.762N_2$$

Component	$0.5CH_4$	$+0.5CO_2$	$+O_2$	→	CO_2	$+H_2O$
Molecular weight	16	44	32		44	18
Mass (kg)	8	+22	+32	=	44	+18

The amount of biogas generated from 1 kg sugar can be calculated using the COD values. Dividing the COD of sugar by the COD of biogas gives: 1.123/1.067=1.053 kg/kg, which suggests that 1 kg of sugar gives 1.053 kg of biogas, the same value as calculated above. This shows that a measurement of COD can be used to calculate the maximum amount of biogas that can be expected from a feed material. The COD of methane is 4 kg/kg (=32/8), so using the same approach, 1 kg of sugar will generate 1.123/4=0.281 kg methane, again in agreement with the figure above.

Calculations for starch

A more complex feed material is starch, a carbohydrate, which is insoluble in water. Starch is a polysaccharide comprising sugar molecules linked together in long chains. Starch is contained in most foods, and is the main component of grains, such as maize, wheat, barley, and millet; and root vegetables, such as potato, cassava, and yam. Starch can be easily hydrolyzed into soluble sugars.

A basic formula for starch is: $(C_6H_{10}O_5)_n$ where n is about 100, giving a molecular weight of about 16,200. Taking the basic equation (dividing everything by 100), starch can be burnt in a stochiometric mix with air and a mass and energy balance defined:

$$C_6H_{10}O_5 + 6O_2 + 22.57N_2 \rightarrow 6CO_2 + 5H_2O + 22.57N_2$$

The gross energy of combustion for food starch is 17,371 kJ/kg (ERCN, 2012), so:

Component	$C_6H_{10}O_5$	$+6O_2$	→	$6CO_2$	$+5H_2O$	Total
Molecular weight	162(00)	32		44	18	
Mass (kg)	162	+192 =		264	+90	
Energy (kJ/kg mole)	−976,108	0		−393,510	−285,830	−2,814,102
Total	−976,108	0	=	−2,361,060	−1,429,150	−2,814,102
Energy (kJ/kg)	−6,025	0		−8,943	−15,879	−17,371

The COD for starch is therefore 1.185 kg/kg (=192/162).

Starch can be fed to an anaerobic digester, where it will be hydrolyzed and then broken down into biogas:

$$C_6H_{10}O_5 + H_2O \rightarrow 3CO_2 + 3CH_4$$

A similar mass and energy balance can be defined.

Component	$C_6H_{10}O_5$	$+H_2O$	→	$3CO_2$	$+3CH_4$	Total
Molecular weight	162(00)	18		44	16	
Mass (kg)	162	+18	=	132	+48	
Energy (kJ/kg mole)	−976,108	−285,830		−393,510	−74,870	
Total	−976,108	−285,830	=	−1,180,530	−224,610	−143,202
Energy (kJ/kg)	−6,025	−15,879		−8,943	−4,679.8	−884.0

The energy released from this reaction is 884.0 kJ/kg starch, which is very similar to that from the digestion of sugar. The reaction is exothermic, but only 5.1% (=884.0/17371) of that produced when starch is completely decomposed to carbon dioxide and water.

For starch, 1 kg produces 1.111 kg of biogas, of which half by volume (0.296 kg) is methane. By volume: 1 kg of starch produces 0.886 m³ of biogas (at NTP) of which 0.443 m³ is methane. Burning the biogas from 1 kg of starch generates

16.487 MJ, so, again, the energy equation balances: $16.487+0.884=17.371$ MJ, which is the energy generated from burning 1 kg of starch in air.

Using the COD value for starch predicts that $1.185/1.067=1.111$ kg of biogas and $1.185/4=0.296$ kg methane.

Simple biogas model

In designing a biogas plant, a simple model is required that can predict the amount of gas that is available from a given amount of feed material, when it is put into the biogas unit. The biogas potential is available for a large number of potential feed materials, but is not useful for predicting the performance of a plant. A simple model that links the gas generated per day with the size of the plant and the amount of material put in each day is provided by the First-Order Rate Model. However, a second parameter, the rate parameter, is required, as well as the biogas potential.

Several biogas design parameters are needed in order to use the model, but most can be calculated from basic measurements.

1. Retention time (R)

   For a batch plant, R is the time the slurry y is left in the digester pit.

   For a continuous digester: $R = \dfrac{V}{v} = \dfrac{\text{Digester pit volume}}{\text{Daily feed volume}}$ days

2. Feed rate

   For a batch plant = Amount fed each loading / time between loading.
   For a continuous plant:

   a) Daily feed volume (v) m³ d⁻¹ $v = \dfrac{w}{\rho_w} + W$, ρ_w = density of feed = 1000 kg m⁻³, W=volume of water added to give a free-flowing slurry.

   b) Daily dry mass feed (m) kg d⁻¹ $v = \dfrac{m}{\text{TS\%} \times 10}$ m³ d⁻¹ (TS = total solids)

   c) Loading rate (r) = daily VS added per m³ digester, (VS=volatile solids or chemical oxygen demand)

   $r = \dfrac{m \times VS\%}{V \times 100}$ kg VS m⁻³ d⁻¹ or kg COD m⁻³ d⁻¹

3. Concentration (S)

   S kg VS m⁻³ or kg COD m⁻³: 'digestible' matter in digester, can be measured in terms of volatile solids or COD

   S_0: initial concentration of digestible matter

For a continuous plant:

$$S_0 = \frac{m \times VS}{v} \quad \text{or} \quad S_0 = \frac{r\,V}{v} \quad \text{or} \quad S_0 = r\,R$$

The First Order Rate model has two basic assumptions (Bulmer *et al.*, 1985).

1. Rate of substrate conversion is directly proportional to substrate concentration.

$$\frac{dS}{dt} = -k\,S \quad \text{where } k = \text{first-order rate constant (d}^{-1}) \text{ and } t = \text{time(d)}$$

2. Volume of gas given off is proportional to mass of substrate digested.

For a batch plant: $G = C\,V\,(S_0 - S)$ G = cumulative gas production (m^3)

For a continuous plant $g = C\,v\,(S_0 - S)$ g = daily gas production (m^3 d^{-1})

C = Biogas potential (m^3 (kg VS)$^{-1}$ or m^3 (kg COD)$^{-1}$)

For a batch digester:

$\frac{dS}{dt} = -k\,S$ can be integrated to give: $S = S_0 \exp\{-k(t - t_0)\}$ where t_0 = lag time, time for digestion to start. Using $G = C\,V\,(S_0 - S)$, gives $G = C\,V\,S_0\left(1 - e^{-k(t - t_0)}\right)$

For a semi-continuous digester:

Mass in-mass out = mass converted/day so:

$$v\,S_0 - v\,S = k\,S\,V \text{ and so: } S_0 - S = k\,S\frac{V}{v} = k\,S\,R,$$

and: $S = \dfrac{S_0}{1 + k\,R}$.

But: $g = C\,v\,(S_0 - S)$, so: $g = C\,v\,S_0\dfrac{k\,R}{1 + k\,R} = C\,V\,S_0\dfrac{k}{1 + k\,R}$

This equation gives a simple link between the gas produced each day, the daily feed volume, the volume of the digester and the concentration of the digestible matter in the feed material.

It does require two other parameters: biogas potential (C) and rate constant (k).

Tests on cattle dung in Reading University gave the following results for batch plants (see Table A1.1) (Fulford, 1988), but different results for semi-continuous plants (see Table A1.2) (Fulford, 1988).

The two different values for each rate constant came from the graphs on which the gas production was plotted. There were two distinct areas of the graph: the first area showed a higher rate constant, but after about 50 days, the rate constant dropped to the lower value. This is taken to indicate that when the more easily digestible material is being processed, the rate is limited

Table A1.1. Parameters for cattle dung in batch plants

Temperature (°C)	Volatile solids		COD	
	Biogas potential C (m³ kgVS⁻¹)	Rate constant k (d⁻¹)	Biogas potential C (m³ kgVS⁻¹)	Rate constant k (d⁻¹)
34	0.895	0.0129 → 0.0033	0.513	0.0188 → 0.0054
25	0.829	0.0068 → 0.0023	0.525	0.0090 → 0.0031
16			0.514	0.0038

by methanogenesis. Once this material has been consumed, the rate is then limited by the hydrolysis of the rest of the feed material.

The consequence of these results is that parameters measured for batch digestion systems (which are usually used to measure biogas potential) cannot be used to predict the performance of semi-continuous plants. Values for biogas potential are available in the literature. Since most of these values are measured using batch digestion, there is a need for measurements of both biogas potential and rate constants for a wide range of biogas feed materials that could be used in semi-continuous plants.

Table A1.2 Parameters for cattle dung in continuous plants

Temperature (°C)	Volatile solids		COD	
	Biogas potential C (m³ kgVS⁻¹)	Rate constant k (d⁻¹)	Biogas potential C (m³ kgVS⁻¹)	Rate constant k (d⁻¹)
34	0.402	0.083	0.347	0.081
25	0.289	0.069	0.237	0.078
16	0.178	0.033	0.164	0.026

Values from batch plant tests on food wastes have also been determined (see Table A1.3) (Banks and Heaven, 2013). The researchers at Southampton University also found two rate constants, one for short retention times, when easily digested material is processed, and another for long retention times, when less digestible material is processed.

Table A1.3 Parameters for food wastes in batch plants

Biogas potential C (m³ kg VS⁻¹)	0.47		
Rate constant k (d⁻¹) (k_1, k_2)	0.73	1.02	0.06

Table A1.4 Properties of dung from various animals (Fulford, 1988)

Animal	Dung per day (kg)	TS %	VS of DM %
Buffalo	14.0	16–20	77
Cow	10.0	16–21	77
Pig	5.0	25	80
100 hens	7.5	48	77
Human	0.2	15–20	90

Note: The data is fairly old, but can be used unless local data is available

Table A1.5 Properties of food wastes

Animal	TS %	VS of DM %	Reference
Domestic organic wastes	41	64	Average of Dutch data (ERCN, 2012)
Food residues	10	80	(Biosantech et al., 2013)

Note: The properties of food waste depend on the contents; local data should be used

---

Example A1.1 Gas production from animal dung

The first order rate model can be used to provide an estimate of the gas production from a biogas plant. Using a GGC biogas plant with a working volume (V) of 6 m^3, fed with dung from 2 cattle and 2 buffalo, the wet mass of daily feed is 48 kg (using data from Table A1.4).

The dry mass is 48 × 18%=8.64 kg and the volatile solids content is 8.64 × 77%=6.65 kg

Normally, the dung is mixed with the same amount of water. Assuming a density for dung of 1000 kg m^{-3}, the daily feed volume (v) is 0.096 m^3. The concentration of volatile solids in the feed material is therefore $S_0 = 6.65/0.096 = 69.3$ kgVS m^{-3}.

$$g = C V S_0 \frac{k}{1+kR} \qquad\qquad R = \frac{V}{v} = 6/0.096 = 62.5 \text{ days.}$$

Using the values of C and k from Table A1.3:

$$g = 0.402 \times 6 \times 69.3 \times \frac{0.083}{1+0.083 \times 62.5} = 2.24 \text{ m}^3 \text{ of biogas a day.}$$

Example A1.2 Gas production from food waste

Using a Khadi and Village Industries Commission (KVIC) biogas plant with a working volume (V) of 34 m^3, fed with 300 kg of wet food waste a day: using the data from the first line of Table A1.5:

The dry mass is 300 × 41%=123 kg and the volatile solids content is 123 × 64%=78.7 kg.

Mixing the food waste 1:4 with water and assuming a density of 1000 kg m^{-3}, the daily feed volume (v) is 1.2 m^3. The concentration of volatile solids in the feed material is therefore $S_0 = 78.7/1.2 = 65.6$ kgVS m^{-3}.

$$g = C V S_0 \frac{k}{1+kR} \cdot R = \frac{V}{v} = 34/1.2 = 28.3 \text{ days.}$$

Using the values of C and k from Table A1.2 for 34 °C:

$$g = 0.47 \times 34 \times 65.6 \times \frac{0.73}{1+0.73 \times 28.3} = 35.3 \text{ m}^3 \text{ of biogas a day.}$$

Note: using a single digester for food waste suggests the food waste should be ground up before it is mixed with water to form a slurry and put in the digester.

Endnotes

a The gross enthalpy of combustion assumes the water that is generated is in liquid form. The net enthalpy of combustion assumes that the water remains as steam.

b The enthalpy of formation of liquid water includes the enthalpy of vaporization, while that for vapour does not. The analysis will also work if the net enthalpy of combustion is used along with the enthalpy of formation of water vapour.

c Normal temperature (20 °C or 293 K) and pressure (1 atmosphere).

References

Banks, C.J. and Heaven, S. (2013) '6-Optimisation of biogas yields from anaerobic digestion by feedstock type', in A. Wellinger, J. Murphy and D. Baxter (eds), The Biogas Handbook, pp. 131–165, Woodhead Publishing Cambridge, UK. <http://dx.doi.org/10.1533/9780857097415.1.131> [accessed 21 July 2014].

Biosantech, T.A.S., Rutz, D., Janssen, R. and Drosg, B. (2013) '2-Biomass resources for biogas production', in A. Wellinger, J. Murphy, and D. Baxter, (eds). The Biogas Handbook, pp. 19–51, Woodhead Publishing Cambridge, UK. <http://dx.doi.org/10.1533/9780857097415.1.19> [accessed 21 July 2014].

Bulmer, A., Finlay, J., Fulford, D.J. and Lau-Wong, M.M. (1985) Biogas: Challenges and experience from Nepal Vols. I and II, Butwal, Nepal: United Mission to Nepal. <www.kingdombio.com/Biogas-vol-I.pdf> [accessed 21 July 2014].

Buswell, A. and Neave, S.L. (1930) Laboratory Studies of Sludge Digestion, Illinois: Water Survey of the State of Illinois. <webh2o.sws.uiuc.edu/pubdoc/B/ISWSB-32.pdf> [accessed 23 July 2014].APPENDIX 1

APPENDIX 2
Biogas plant design details

Abstract

Various designs of a biogas plant have been developed by the different biogas extension projects. Since these plants have been designed to do similar jobs, basic dimensions are similar. Earlier designs used longer retention time, and therefore larger volumes. The required retention time is affected by the local ambient temperature.

The mathematical formulae required to calculate various volumes are given. Actual measurements may be slightly different from those calculated from the formulae, as the shapes produced when plants are built may not fit exactly to the shapes defined by the formulae.

Keywords: biogas, anaerobic digestion, chemistry, thermodynamics

Design considerations

The dimensions of domestic biogas plants built by various groups are roughly similar, especially if they use a similar feed material, such as animal dung, and are designed to supply sufficient gas for a family to cook their food. However, there has been a trend over time to reduce the working volume, thus reducing the hydraulic retention time and the plant cost. The typical working volume of the early floating drum designs was 7 m^3, giving a retention time of almost 60 days when fed with dung from six cattle. Deenbandhu plants built more recently only have a working volume of 2 m^3, with a retention time of less than 20 days for the same feed.

The retention time also depends on the local ambient temperature. The earlier plants were built in northern India, where the temperature is lower, especially in winter. More recent programmes in India have been based in the south, which has a more consistent temperature throughout the year.

Steel-drum plant sizes

The original floating drum designs developed by Khadi and Villages Industries Commission (KVIC) had a main digester tank size of 15 m^3 (Lichman, 1983). The systems developed in Nepal, based on the KVIC designs, had a range of sizes, as listed in Table A2.1 (Bulmer *et al.*, 1985), which lists retention time and expected gas production (if the ambient temperature is optimal). Most

Table A2.1 Standard sizes of floating drum plant used in Nepal (Bulmer *et al.*, 1985)

Pit volume (m^3)	Drum volume (m^3)	Dung (kg/ day)	Retention time (day)	Gas (m^3/ day)	Gas (ft^3/ day)	Gas nominal (ft^3/day)
7.1	1.7	60	59	2.5	88	100
13.0	3.4	120	54	5.6	198	200
24.3	6.0	210	58	9.9	350	350
34.0	8.5	300	57	14.2	501	500

plants were built to the smallest (standard) size. The larger sizes were used to generate gas to drive engines, for food milling or for irrigation, but required dung from many more cattle. The original description used the nominal daily gas production in ft^3 per day. The gas production figures in m^3 per day were calculated, using a first-order rate model at an optimum ambient temperature.

The depth of the main digester pits that were built in several areas of Nepal had to be reduced, because the underground water table was fairly high during some seasons of the year. The tapered design of pit was developed had a much larger diameter hole at the base, which tapered to a diameter that matched the appropriate gas drum near the surface of the ground. The total volume of this tapered hole was calculated to be the same as for a straight hole, so the performance would be the same for both designs of a standard size. The tapered design used a wall across the base of the digester pit, between the inlet and outlet pipes, to prevent dung flowing from one pipe to the other.

Fixed-dome plant sizes, as used by BSP in Nepal

The original fixed-dome biogas plants were developed by Development and Consulting Services (DCS), but the designs were simplified by the Gobar Gas Company (GGC). The 'standard' design, called the GGC 2047, was defined

Table A2.3 Standard sizes of fixed-dome plant used in Nepal (Devkota, 2001)

SN	Capacity (m^3)	Daily dung feed (kg)		Number of cattle required
		Hills	Terai	
1	4	24	30	2–3
2	6	36	45	3–4
3	8	48	60	4–6
4	10	60	75	6–9
5	15	90	110	9–14
6	20	120	150	14+

Table A2.4 Volumes used for fixed-dome plants as used in Nepal (Devkota, 2001)

Nominal size (m³)	Digester (m³)	Gas storage (m³)	Retention time (day)	Gas production (m³/day)
4	2.8	1.2	80	1.4
6	4.4	1.7	75	1.8
8	5.8	2.2	83	2.2
10	7.5	2.8	83	3.1
15	11.4	3.9	83	4.2
20	14.2	5.8	83	6.4

for six basic sizes, which are listed in Table A2.3. The ambient temperature in the hills of Nepal is lower than that on the plains area (Terai), so more dung is required to achieve the designed gas production.

The actual volumes of different parts of the plants are given in Table A2.4. The long retention times are used to ensure that users in the hills of Nepal, with lower ambient temperatures, still get good gas production.

Deenbandhu plant sizes, as used by VK NARDEP (a South Indian NGO) in South India

The standard sizes for Deenbandhu plants, as built in South India are given in Table A2.5.

Table A2.5 Volumes used for Deenbandhu plants in South India (Satyamoorty, 1999)

Size	Dung kg day⁻¹	Number of cattle	To cook for people	Digester volume m³	Gas store m³	Total volume m³
1	25	2–3	3–4	2.04	0.50	2.54
2	50	4–6	5–8	4.05	0.67	4.72
3	75	7–9	9–12	6.09	0.97	7.06
4	100	9–11	12–16	7.99	1.37	9.36

Dimensions of floating drum biogas plants

A technical drawing of a straight biogas plant with a floating drum used in Nepal is provided in Figure A2.1. The key dimensions are shown in Table A2.6.

A technical drawing of a tapered biogas plant with a floating drum as used in Nepal is provided in Figure A2.2. The key dimensions are shown in Table A2.7.

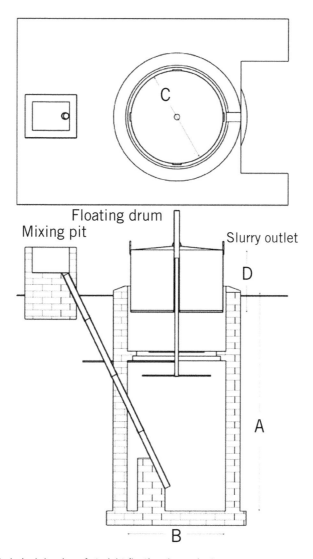

Figure A2.1 Technical drawing of straight floating drum plant

Table A2.6 Dimensions for straight floating drum plants used in Nepal (Bulmer *et al.*, 1985)

Dimension (m)	Plant volume (m³)			
	7.10	13.00	24.00	34.00
A	3.62	4.63	4.93	5.54
B	1.60	2.00	2.60	3.59
C	1.50	1.88	2.50	2.80
D	1.00	1.22	1.22	1.40

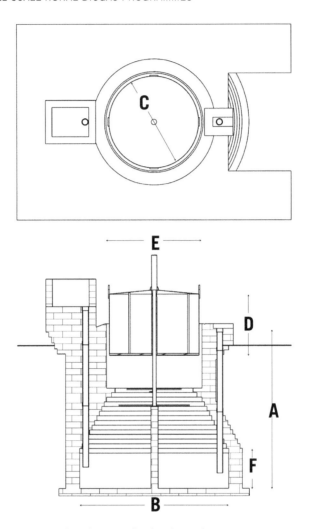

Figure A2.2 Technical drawing of tapered floating drum plant

Table A2.7 Dimensions for tapered floating drum plants as used in Nepal (Bulmer *et al.*, 1985)

Dimension (m)	Plant volume (m³)			
	7.10	13.00	24.00	34.00
A	2.52	3.09	3.30	3.77
B	2.50	2.90	3.90	4.23
C	1.50	1.88	2.50	2.80
D	1.00	1.22	1.22	1.40
E	1.60	2.00	2.60	2.90
F	0.58	0.91	0.81	1.05

Table A2.8 Dimensions for GGC dome plants as used in Nepal (Devkota, 2001)

Dimension (m)	Plant volume (m^3)					
	4.00	6.00	8.00	10.00	15.00	20.00
A	1.62	1.76	2.4	2.39	2.77	2.89
B	2.04	2.44	2.7	3.08	3.5	3.98
C	0.86	0.92	1.05	0.94	1.15	1.15
R	1.13	1.44	1.67	1.93	2.36	3.30

Table A2.9 Dimensions for Deenbandhu plants as used in South India (Satyamoorty, 1999; VK NARDEP, 1993)

Dimension (m)	Nominal size			
	1.00	2.00	3.00	4.00
A	0.83	1.05	1.21	1.35
B	2.10	2.55	2.90	3.18
C	1.05	1.28	1.45	1.59
R	1.70	2.02	2.28	2.42

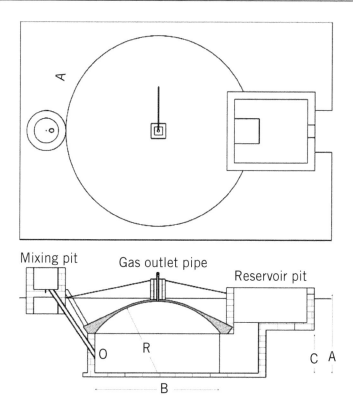

Figure A2.3 Technical drawing of GGC dome plant

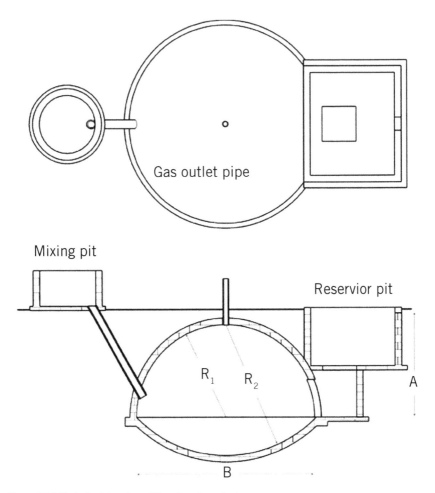

Figure A2.4 Technical drawing of Deenbandhu plant

Equations to calculate plant volumes

The internal volume (V) of a cylinder, such as the digester pit of a floating drum biogas plant, can be calculated, where D is the diameter of the cylinder and H is the depth:

$$V = \frac{\pi}{4} D^2 H$$

The gas storage volume (V) of an underground biogas plant is a segment of a dome, or a cap cut from the top of a dome of radius (R). The horizontal radius (r) of the base of the cap is related to the height (h) of the cap by:

$$R^2 = r^2 + (R - h)^2$$

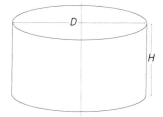

Figure A2.5 Cylinder showing dimensions

Which gives:

$$h = R \pm \sqrt{R^2 - r^2}$$

Now:

$$V = \frac{\pi}{3} h^2 (3R - h)$$

Rearranging gives:

$$V = \frac{\pi}{6} h (3r^2 + h^2)$$

Example A2.1 Volume of floating drum plant

Taking the dimensions of a straight biogas pit: $D=2.44$ m (B) and $H=0.92$ m (A) gives

$V = \frac{\pi}{4} 2.44^2 \times 0.92 = 4.3$ m³. The actual figure, 7.1 m³, is the result of extra masonry (ledge and inlet pipe support) occupying some of the volume.

The volume of the gas drum is: $V = \frac{\pi}{4} 1.5^2 \times 1.0 = 1.77$ m³. The quoted volume (1.7 m³) allows for steel structures inside the drum, including scum breaker bars.

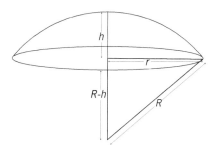

Figure A2.6 Sector of sphere showing dimensions

Example A2.2 Volume of GGC dome plant

Taking the dimensions of the biogas pit: $D=2.44$ m (B) and $H=0.92$ m (C) gives:

$V = \dfrac{\pi}{4}2.44^2 \times 0.92 = 4.30$ m^3. The volume of the gas dome is given by $R=1.44$.m (R) and $r=1.22$ m $(B/2)$, so $h = 1.44 \pm \sqrt{1.44^2 - 1.22^2} = 0.68$ m and $V = \dfrac{\pi}{6} \times 0.68 \times \left(3 \times 1.22^2 + 0.68^2\right) = 1.74$ m^3.

The total volume is: $4.30 + 1.74 = 6.04$ m^3, very close to the nominal value of 6 m^3.

Example A2.3 Volume of Deenbandhu dome plant

Taking the dimensions of the bottom spherical section: $r=1.28$ m $(B/2)$ and $R_2=0.92$ m gives: $h = 1.28 - \sqrt{1.28^2 - 0.92^2} = 0.45$ m and $V = \dfrac{\pi}{6} \times 0.45 \times \left(3 \times 1.28^2 + 0.45^2\right) = 1.21$ m^3

For the main dome $h=1.28$ m (the dome is a complete hemisphere), so:

$V = \dfrac{\pi}{6} \times 1.28 \times \left(3 \times 1.28^2 + 1.28^2\right) = 4.34$ m^3. The total volume is: $4.34 + 1.21 = 5.55$ m^3.

Not all of the main dome is used, so the quoted volume is 4.72 m^3 (Satyamoorty, 1999)

References

Bulmer, A., Finlay, J., Fulford, D.J. and Lau-Wong, M.M. (1985) Biogas: Challenges and experience from Nepal Vols. I and II, Butwal, Nepal: United Mission to Nepal. <www.kingdombio.com/Biogas-vol-I.pdf> [accessed 17 July 2014].

Devkota, G.P. (2001) Biogas Technology in Nepal: A sustainable source of energy for rural people, Kathmandu: Mrs. Bindu Devkota, Maipee.

Lichtman, R.J. (1983) Biogas Systems in India, Arlington: Volunteers in Technical Assistance. <www.build-a-biogas-plant.com/downloads/assets/Biogas Digesters in India.pdf> [accessed 23 July 2014].

Satyamoorty, K. (1999) Biogas-A boon, handbook on biogas, Kanyakumari, South India: Vivekananda Kendra – Nardep. <www.vknardep.org/publications/english-books/189-biogas-a-boon.html> [accessed 23 July 2014].

VK NARDEP (1993) Biogas: A manual on repair and maintenance, Kanyakumari: Vivekananda Kendra, Natural Resources Development Project. <www.vknardep.org/publications/english-books/197-biogas> [accessed 23 July 2014].

ERCN, Energy Research Centre of the Netherlands (2012) 'Phyllis, database for biomass and waste' [website-database] <www.ecn.nl/phyllis> [accessed 21 July 2014].

Fulford, D. (1988) Running a Biogas Programme: A handbook, London: Practical Action Publications (Intermediate Technology Publications). <http://developmentbookshop.com/running-a-biogas-programme-pb> [accessed 21 July 2014].

NIST (2011) 'NIST Chemistry WebBook' [website-database] <http://webbook.nist.gov> [accessed 21 July 2014].

APPENDIX 3
Building a masonry biogas plant

Abstract

Biogas units made in large numbers in Asia are made from masonry by manual labour, using cylindrical or spherical shapes. The basic building techniques are defined and use basic skills that need to be gained by the masons who are doing the work. Similar approaches need to be used for the different designs, although there are particular approaches required for each different design.

Keywords: biogas, anaerobic digestion, chemistry, thermodynamics

Basic design concepts

The biogas plants used for domestic purposes in Asia and Africa are mainly made from brick and/or concrete, to reduce overall costs and also to ensure reliability and long lifetime (see Chapter 6). The building techniques are designed to be simple, so masons with limited education can use them. Many masons can develop very high skills as they gain experience in building these designs.

One feature of building these designs is the use of manual labour, rather than machinery, to dig holes. This allows the earth around a biogas plant to be accurately shaped and the surrounding earth is often used to support the structure of the plant. Some versions of these designs use dug earth as a mould over which to cast concrete; this reduces the need for specially designed wood or steel moulds, which can be expensive and difficult to transport.

Another feature is the use of cylindrical and spherical shapes for the masonry. Such shapes can be cast in concrete using suitable moulds. Bricks are rectangular in shape, but can be used to make curved shapes by placing each one at an angle to the previous ones. Extra mortar is needed to fill in the gaps. The minimum space between two bricks should be 12 mm. The mortar mix should be as dry as possible, and the bricks should be soaked in water before they are laid.

Sufficient building materials and tools need to be available at the site before the work starts. A datum level should be marked so that depths can be measured during excavation and building. The materials used for a masonry work need to be of good quality (Kudaravalli 2014a; Bulmer *et al. 1985*). If bricks are used, they need to be properly fired and not be porous or easily crushed. Sand and aggregate (gravel) should be free from mud and other contamination. They can be washed if necessary. If stone is used, it needs to be strong and not liable to crumble. Ordinary Portland cement is used, but it must be fresh and dry.

Main digester pit

A biogas plant consists of the main digester pit, with floor and walls; a gas storage volume, either a floating drum or a fixed-dome; a slurry inlet; a slurry outlet, with reservoir pit, if it is a displacement digester; and a gas outlet (Kudaravalli 2014b). Space must be allowed for the building of a latrine, if one is to be attached to the plant.

Most of these plant designs are made underground, therefore the first step in building a plant is to clear a patch of ground that is large enough in which to fit the size of plant that is chosen. The area should be levelled, as far as possible, and the positions of the different parts of the plant marked out with pegs. The main pit is usually cylindrical in shape, so a circle is marked out on the ground, using a central peg and a string. If the soil is reasonably firm, the hole can be dug with vertical walls. If the soil is weaker, the walls may need to slope outwards, to prevent them collapsing.

When the hole has reached a suitable depth, the floor of the digester pit can be made. Different designs use either a flat floor or one that is concave. If a concave floor is required, its shape is usually made according to a metal template that is rotated around a central peg. The template is moved downwards as soil from the centre is removed. A concave floor allows a larger volume for the main digester pit for a given wall depth. A flat floor is easier to make, because the edges of the pit are excavated to the same depth as the centre. In structural terms, there is little difference between a flat and a concave floor, in both the weight of the slurry in the digester pit is supported by the undisturbed soil under the floor.

The floor can be made from either poured concrete or bricks. If bricks are used, they can be laid in rings around a central group, with cement mortar (a mix of 1:6 cement to sand) between them. Concrete floors should be made to a uniform thickness of 75 mm for the smaller plants or 100 mm for larger plants. A thicker layer (up to 150 mm) of concrete can be used for the edge of floor, where it acts as a foundation for the plant walls. The floor should extend at least 75 mm beyond the outside of the wall. A mix of 1:2:4 of cement, sand, and aggregate is strong enough for a floor. Most of the plants built in the large programmes do not use steel reinforcement in the floor. If the soil below the floor is weak, it should be well compacted before the floor is made. If the ground is of very poor quality, a layer of rubble, followed by a layer of compacted sand can be used as a base over which the floor can be poured.

Once the concrete or mortar used in the floor has set, the walls can be built. In the case of floating drum and Janata or Gobar Gas Company (GGC) design plants, the walls are made as a cylinder of brickwork or stonework, held together with cement mortar. The radius of each course of bricks or stones in the cylinder is measured with a stick or a piece of rope of fixed length from a central pole that has been carefully set vertically on the centre of the floor. The top of this pole is supported with guy ropes and pegs to ensure it is fixed in place. A length of steel gas pipe can be used as this datum pole; it is removed when the walls are complete. As the courses are built up, the space between the bricks or stones and the sides of the hole should be backfilled with earth, which is rammed in to support the walls.

In the case of the Deenbandhu design of plant, the walls are built using bricks to form a dome, with a centre placed at the level of the edge of the floor. A brick plinth is built to the right height, with a short piece of steel rod fitted in the top, to act as the datum for the centre. A length of rod or rope is used to measure the radius from this central datum point and each course of bricks is placed so the inside edge is at an equal distance from this datum. As the walls are built up, they curve inwards towards a central point. As new bricks are laid, they can be temporally held in place with wire clamps, or poles, on which other bricks are placed to act as weights (see Chapter 6, Figure 6.7). Once a whole course of bricks is in place, they should form a ring in which each brick supports the others. It is important to keep the thickness of the mortar between the bricks as uniform as possible, with a minimum spacing of 12 mm.

Inlet and outlet

As the walls of any design are built, pipes or chambers must be included to allow slurry to enter and leave the main digester pit. Most designs use a pipe through which the slurry enters the digester. The type of pipe used locally for making land drains is the most convenient to use, preferably made from cement, which bonds well into the digester wall. Plastic pipe can be used if cement pipe is not available. The inlet pipe should be straight, so it can be cleaned with a rod if it becomes blocked. Different groups suggest different diameters, from 100 mm to 200 mm, for the inlet pipe. Larger diameter pipes are required if the slurry is not well mixed before it is added.

The slurry outlet can also be via a similar pipe, if access to the digester pit is possible from the top. Some designs of floating drum plant use a vertical wall across the width of the pit, with inlet and outlet pipes either side. The slurry is then forced to flow up and over the wall before it leaves, preventing hydraulic short-circuit. Other floating drum designs are made with a deep pit, so the slurry enters at the base and leaves via an opening in the top of the wall.

Access to a fixed-dome plant digester can either be via the slurry exit or by having a manhole in the top of the dome, closed by a removable cover. If access is via the slurry exit, a chamber that has horizontal dimensions of at least 600 × 600 mm needs to be built alongside the digester wall, with an opening that allows the slurry to enter this chamber. The chamber can have a square or a circular cross-section and is built of the same materials as the main digester pit. The floor of this separate chamber can be at the same level as the floor of the main digester or higher. If the outlet chamber is used for access during the building of the plant, the floor needs to be placed to allow masons to get in and out easily. The top of the outlet chamber is the base of the slurry reservoir.

Floating drum plant

The top of a floating drum plant needs to include a system to support the steel drum in a vertical orientation, so that it can move easily up and down as it fills with gas. The drum needs to have a ledge on which to sit when it is empty of gas. This ledge also diverts biogas formed in the digester, so that it enters the metal tank and does not leak out of the gap between the tank and the digester walls.

The Khadi and Village Industires Commission (KVIC) floating drum design is made with a central pipe that can fit over a vertical guide pipe mounted in the main digester pit. This guide pipe is supported by metal frames that are mounted into the digester wall as it is built. These frames should be adjusted so the guide pipe is vertical and the drum will move up and down freely. The ledge is made using bricks arranged so they extend inwards from the wall into the digester.

Deenbandhu plant with manhole

The top of a Deenbandhu plant is formed as the walls are built inwards, following the spherical shape. It is recommended that a 'strong ring' be included in the wall (Koottatep, Ompont, *et al.* 2004; Sasse, Kellner, *et al.* 1991) about halfway up its height. This ring supports the weight of the dome above it, especially the tensile forces that might cause the dome to bulge outwards at this point. The wall of this ring is made thicker by placing the bricks at right angles to their normal position; the ends protrude from the wall. After the brick layers above have been put in place, a concrete ring is poured that fills in the space above this thicker brick layer. If the surrounding soil is weak, a steel reinforcing-rod can be placed in the concrete of the strong ring to give it additional strength.

If the plant is to be made with a manhole in the top, the dome is continued until a hole of around 650 mm is formed (large enough for a man to climb through). A vertical cylinder of brickwork, 'the neck', is made around this hole, so that a concrete plug can be placed inside it. The concrete plug is made in a tapered mould and usually includes a gas outlet pipe and lifting handles (NSPRC 1984). An alternative approach is to include the gas outlet pipe in the neck, below where the concrete plug will fit. The inside of the neck is plastered with mortar, and an internal taper is made to fit the concrete plug. If the tapered mould is made from a sheet of steel, the same mould can be used for both the plug and the inside taper, to ensure they both have the same shape. There should be a 20 to 25 mm gap between the plug and the inside taper that can be filled with wet mud to act as a seal. The plug can be held in place with three wedges that fit into holes on the inside of the taper. The neck above the taper is filled with water to keep this mud seal moist. A cover (usually made of concrete reinforced with steel) is placed over the top of the neck to reduce the rate at which this water evaporates. However, this water must be kept topped up regularly to ensure the mud seal does not dry out. Oil can also be added to reduce evaporation.

If the manhole cover needs to be removed to access the inside of the plant, it must be replaced very carefully, so that a new seal can be made. The old mud should be carefully washed off and new mud obtained to make a new seal between the plug and the tapered hole.

Deenbandhu plant without manhole

If the access to the plant has been designed to be through the slurry exit, the dome can be continued until it can be closed at the top with bricks and mortar. Bricks may need to be cut to shape to fill the remaining space. A piece of wood or metal on a support frame will need to be placed under the last few rings of bricks, to prevent them falling out of the top of the dome until the mortar sets. The gas outlet pipe must be fitted into the top of the dome, at the highest point.

DCS and Janata plants

The Gobar Gas Company (GGC) and Janata designs use a mould to form the dome. The dome can either be made of concrete poured over the mould, or made using concrete blocks placed on the mould with cement mortar in between. The Janata design uses a much shallower spherical shape than the Deenbandhu design, so it is difficult to get bricks or blocks to form self-supporting rings.

The GGC design uses a mud mould, made by filling the digester pit with earth and shaping the top with a metal template that is rotated around a central post. The dome is then cast in concrete; the outside is shaped by a second template, which is rotated around the gas outlet pipe, which is placed at the highest point of the dome. The shape of the dome is designed so the whole structure is in compression and the weight is taken by the earth outside the main walls. The outside edges of the dome are made much thicker (e.g. 300 mm) than the middle (e.g. 60 mm) to act as a foundation to support this weight. Once the concrete has become hard, the earth is removed from under the dome, via the access hole that leads to the slurry reservoir at the side of the main digester pit.

The Janata design can use a similar earth mould, or use one made from steel or wood. Steel moulds are usually designed to be easily dismantled, so they can be used many times. The dome is made up from a set of metal plates that are each curved to form a section of a sphere. Ideally the plates should be curved in two dimensions, similar to segments of an orange, but this makes them difficult to make. The length of each plate can be curved, while the width is straight. These plates are fairly complex to design; the edges must be curved when they are cut out of a flat plate to ensure they fit together correctly after they have been curved to shape. In order for these plates to be stiff enough to take the weight of the concrete, each edge has a rib welded at right angles to the surface. Bolts or pins are used to hold these ribs together when the mould is constructed. The system must be carefully designed so that the sections will come apart, even when they have the weight of the concrete on top of them. Spacers can be placed between the sections, so that when they are removed, the sections can be taken away from the set concrete more easily.

The process of making a curved plate in wood requires the use of steamed plywood as used in boat building, which is not easy to do. Wood is usually made into a platform on which earth can be piled to form the surface of the mould. In India, the problem of making moulds of the correct shape resulted in the Janata plants having a higher failure rate than the Deenbandhu design.

Plastering and sealing masonry

Most brickwork and concrete is porous unless it is sealed. This is usually done by applying cement plaster that fills in the holes of the bricks or concrete. The easier areas of masonry to seal are those that are always under the surface of the slurry in a digester. If slurry does leak through the pores of the brick or concrete, it will dry within those pores and be effectively self-sealing. However, there is a danger that slurry is lost through the porous surface, or cracks in it, and this will pollute the ground around the plant.

The more complex issue is the leakage of biogas though porous surfaces or cracks. A cement plaster layer will cover the worst cracks, but is slightly porous itself. There are several suggested solutions. A cement 'wash', made from cement mixed with water, can be used. Several coatings can be applied. In Nepal, acrylic emulsion paint is mixed with the final plaster coating and also with the cement wash (Fulford 1988). When the paint sets, it fills in the pores of the cement and gives a good seal. A report from China suggested the use of vinyl emulsion paint, which does a similar thing (Tang, Xie, *et al.* 1985).

Alternative approaches

If the soil is very weak, it is possible to use a technique that was developed for digging wells. Pre-cast concrete rings are made to the correct diameter for the plant design. A ring is placed on the ground and the soil excavated from inside, so the ring sinks into the soil. Further rings are placed on top, as the hole is dug, until the base of the bottom ring is at the right depth. The floor can then be made from concrete poured at the base of the lower ring.

A technique suggested for use in China suggests that a circular trench is dug to act as an earth mould for the main digester walls. Inlet and outlet pipes or chambers can be put in place. Concrete is then poured into the trench, up to the correct height for the walls and allowed to harden. The dome can also be made of poured concrete by shaping the earth inside the trench; either at the same time or after the concrete in the walls has hardened. Once the concrete of the dome has hardened, the earth inside can be removed, via a manhole in the top of the dome or an access chamber at the side. The floor can then be made, again using an earth mould over which concrete is poured.

References

Bulmer, A., Finlay, J., Fulford, D.J. and Lau-Wong, M.M. (1985) Biogas: Challenges and experience from Nepal Vols. I and II, Butwal, Nepal: United Mission to Nepal. <www.kingdombio.com/Biogas-vol-I.pdf> [accessed 17 July 2014].

Fulford, D. (1988) Running a Biogas Programme: A handbook, London: Practical Action Publications (Intermediate Technology Publications). <http://developmentbookshop.com/running-a-biogas-programme-pb> [accessed 21 July 2014].

Koottatep, S., Ompont, M. and Tay, J.H. (2004) Biogas: A GP option for community development, Asian Productivity Organisation, Tokyo <www.apo-tokyo.org/gp/51_6biogasmain.htm> [accessed 17 July 2014].

Kudaravalli, K.K. (2014a) Construction Manual for Biogas Technology, Kolar, South India: SKG Sangha. <http://foundationskgsangha.org/Construction Manual.pdf> [accessed 17 July 2014].

Kudaravalli, K.K. (2014b) Extension Manual for Biogas Technology, Kolar, South India: SKG Sangha. <http://foundationskgsangha.org/Extension Manual.pdf> [accessed 17 July 2014].

NSPRC, N.S. of the P.R. of C. (1984) The Collection of Designs for Household Hydraulic Biogas Digesters in Rural Areas, China State Bureau of Standardization, Beijing.

Sasse, L., Kellner, C. and Kimaro, A. (1991) Improved Biogas Unit for Developing Countries, Eschborn: GATE, GTZ Germany. <http://www2.gtz.de/Dokumente/oe44/ecosan/en-improved- biogas-unit-1991.pdf> [accessed 17 July 2014].

Tang, Z.G., Xie, X.U. and Wu, D.C. (1985) 'Study on polymer seal paint of concrete biogas digester', in Proceedings of the Fourth Annual Symposium on Anaerobic Digestion. Guangzhua, China: China State Biogas Association.

APPENDIX 4
Basic gas pipe fitting

Abstract

The biogas from a biogas plant is only beneficial if it can be piped to where it is needed and used efficiently. Gas leaks from piping often cause operators to decide that their biogas plants are not working. Biogas is being produced, but it is lost through the leaks. Biogas contains water vapour, which can condense in pipes and block them, so measures need to be designed for the water to be removed. As gas flows down a pipe, friction between the gas and the walls of the pipe causes pressure losses, which can be calculated.

Keywords: biogas, anaerobic digestion, chemistry, thermodynamics

Gas pipes

While gas pipes can be made of a range of different materials, there are some basic approaches that will ensure pipes are leak tight and transfer gas effectively. Gas pipes and fittings can be made of a range of materials, such as metals and plastics. There are defined standard sizes for pipes, but the details can be complex. The factor that influences gas flow is the inside diameter of the pipe, more gas can flow along larger pipes. Pipes are designed for a range of pressures, so the wall thickness depends on both the internal pressure and the material of the pipe. Pipes tend to be defined by the outside diameter, as gas fittings usually fit on the outside of pipes.

Biogas is usually at low pressure so 'standard' pipes are used, which are usually defined as 'Schedule 40'. These pipes have thicker walls than are strictly necessary, but are the ones that are usually easily available. Table A4.1 gives a list of the sizes of typical pipes that are used for biogas. The standard sizes

Table A4.1 Typical pipe sizes for gas pipes (The Engineering Toolbox, 2010)

Nominal size (mm)	Nominal size (inch)	Typical outside diameter (mm)	Typical inside diameter (mm)
15	0.5	21.34	15.80
20	0.75	26.67	20.93
25	1	33.40	26.65
32	1.25	42.16	35.05
40	1.5	48.26	40.89
50	2	60.33	52.50

are defined for steel pipes. Actual pipes that are available could have different sizes, especially those made from other materials.

Pipe fittings are of several different types: screwed, welded, compression or pushed. A screwed fitting uses a screw thread on the pipe that is screwed into a similar thread on the fitting. The pipe wall must be thick enough to take the thread. Welded joints are used when the pipes are to be permanently fixed together. Pipes can be 'butt welded' together, where the ends of each pipe are welded directly to each other, or the pipe ends can be welded into fittings. Compression fittings are used when the pipes need to be disassembled. Compression fittings are designed to use a seal that is compressed onto the surface of the pipe by tightening a nut onto a screw thread on the fitting. The seal can be a simple rubber 'O' ring or be specially designed for the fitting. Flexible pipes can be pushed over pipe fittings that are usually ribbed to grip the inside of the pipe. Clamps can be used to grip the outside of the pipe and compress it onto the fitting.

Suppliers of piping usually sell a piping system that includes the pipe, fittings suitable for use with the pipe, and the tools required to use the fittings. It is important to purchase parts that all belong to the same piping system; parts are seldom interchangeable. Some piping systems use pipe sizes that do not meet the standards listed in Table A4.1, so are even less likely to match other piping systems.

The key to ensuring a gas fitting is leak tight is cleanliness. All the parts of a fitting must be carefully cleaned because dirt can cause leaks. Many fittings can be used with sealing materials, such as grease or jointing compounds, to improve the seal. Such compounds should be used sparingly inside the fitting. Attempts to seal a pipe to a fitting by applying such compounds on the outside is usually a futile exercise.

Screwed fittings

Screwed fittings are usually used in steel pipes, which are much stronger than plastic pipes. The classic steel pipes used for gas are GI pipes, steel coated with zinc (galvanising), which are connected together with threaded fittings (see Figure A4.1). The pipes are usually supplied in lengths of 2 m or 3 m, so straight connectors are used to make longer lengths. Fittings are usually supplied with a suitable internal thread for fixed sizes of pipe (such as those defined in Table A4.1). These are usually BSP (British Standard Pipe) threads, although some places use NPT (National Pipe Thread), which was defined in the USA. These threads can be either straight or tapered. Tapered threads become tighter as the pipe is screwed into the fitting, but are more difficult to form. The ends of the steel pipes are threaded using dies that use the same standard as the fittings.

The layout of the overall gas pipeline can be defined by placing pipes in position and cutting to size any sections that are shorter than full pipe lengths. Threads can be made on the ends of pipes where they are needed. The threads

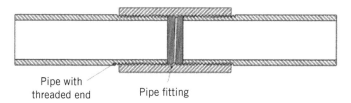

Figure A4.1 Cross-section through threaded pipe fitting

need to be cleaned of swarf (metal fragments) and cutting oil before they are screwed into the sockets in the fittings to check that they work.

When the final fitting is done, the threads need to be sealed. Traditionally the threads were sealed with putty and string, but there is a range of better sealing systems now available. The most effective is PTFE tape, which is wrapped around the thread a few times. When the thread is screwed into the socket, the PTFE plastic distorts to fill in any gaps. Flexible sealing compounds (including PTFE liquid) can also be spread over the threads. They act as lubricants when wet, but dry to form flexible seals. Ideally sealants designed for gas pipes should be used. The pipes are tightened into the sockets using pipe wrenches.

Some plastic pipe systems use screw connectors, but plastic pipe threads are much weaker. Care must be taken when the fittings are tightened. Piping systems that use plastic screw connectors often come with a sealing compound that acts as a lubricant when the pipes and fitting are screwed together, but which sets to form a good gas seal.

Welded pipe fittings

Welding can be used for both metal and plastic pipes. However, it is seldom used for small steel pipes – it involves either high temperatures from oxyacetylene flames or high voltages from arc welding equipment. Copper pipe is sometimes used for gas pipe, although it is usually too expensive for regular use. Copper cannot be welded, but can be brazed or soldered together.

Thicker High Density Polyethylene (HDPE) pipe can be butt welded together. The two ends of the pipe must be cut as squarely (at right angles to the axis of the pipe) as possible. The ends are pressed against the two sides of a heated plate until they begin to melt. The plate is then removed and the pipe ends are pushed together. They must be held together in the correct position until the plastic cools and hardens. Metal clamps are available that help to do this.

Pipe systems that use welded plastic pipe fittings are easier to use. The outside surface of the pipe is welded to the inside surface of the fitting. The heated welding tool has a socket on one side into which the pipe fits and a spigot on the other, over which the socket fits. The pipe needs to be cut squarely and the pipe end made smooth. The basic principle of cleanliness of joints applies. Both the inside of the socket and the surface of the end of the pipe should be carefully cleaned of any contamination (dirt and grease) that could affect the weld.

The welding tool is usually electrically heated, with a thermostat that should accurately control the temperature. The surfaces of the tool that are in contact with the plastic also need to be carefully cleaned before each weld. The fitting is pushed firmly over the spigot and the pipe pushed inside the socket and allowed to reach the correct temperature. The welding tool is then removed and the pipe quickly pushed into the socket on the fitting. The two parts should be held firmly together until the welded surfaces have cooled and solidified.

Different types of plastic have different welding temperatures, so it is important to check that the temperature setting is the correct one. For example, PPR plastics (Polythene-Random, a medium density polyethylene) should be welded at 260 °C ±5 °C. An electronic thermometer, with a thermocouple sensor, can be used to check the welding temperatures.

PVC (polyvinyl chloride) and ABS (acrylonitrile butadiene styrene) pipes cannot be heat welded, but they can be joined by solvent welding. Suitable glues can be purchased for the particular pipe system that is being used. These glues work by dissolving a surface layer of the plastic, which forms a liquid film between two components when they are put together. The film becomes solid as the solvent evaporates away and bonds the surfaces together. The solvents (e.g. acetone) are very volatile and are often flammable, so care should be taken in their use. Solvents can also be poisonous if inhaled excessively, they have strong smells and are irritating to the eyes, nose, and throat; they should be used in a well-ventilated space. The glue container should be kept closed as much as possible and only opened when glue is required.

The pipe needs to be cut squarely (at right angles to the axis of the pipe) and the end smoothed. The surface of the pipe and the inside surface of the fitting need to be cleaned of contaminants (dirt and grease). Some piping systems recommend the use of a primer to prepare both surfaces. The glue is then spread on the pipe and the inside of the fitting, usually with a brush. The parts are then placed together and can be quickly twisted (about 90°) in one or both directions to spread the glue evenly around the surfaces. Excess solvent can be wiped off. The fitting must then be held firmly together while the solvent evaporates (about 30 seconds).

If PVC is to be heat welded, it requires high-frequency heating machines. This technique is normally used for welding sheets of material and is seldom used for making pipe connections.

Compression fittings

Compression fittings were initially designed for copper pipes. A brass compression ring (ferrule) is pushed over the end of the pipe in front of a compression nut. The end of the pipe is then pushed into the brass fitting and the compression nut is then screwed onto the thread on the outside of the fitting (see Figure A4.2). The ferrule is squeezed into the shape at the end of the fitting by the shaped nut, so that it grips the copper pipe tightly, distorting it slightly to give a good seal. The nut has to be turned by a spanner (wrench) to provide sufficient force to distort the ferrule to the correct shape. The advantage of this type of fitting is that the nut can be undone and the pipe removed. If this is done too many times, the sealing surfaces are distorted and become less effective.

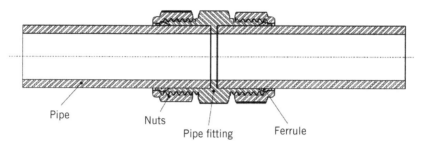

Figure A4.2 Cross - section through compression pipe fitting

The key to making a gas-tight fitting is, again, cleanliness. The end of the pipe should be cut squarely and cleaned of contamination (dirt and grease). The inside of the fitting and the nut should also be cleaned. A well-made fitting should give a gas-tight seal. However, the seal can be enhanced with a thin layer of high viscosity grease or flexible sealant between the ferrule and the seating face of the fitting. Another key feature of such fittings is that the nut should be tightened carefully. Once the nut is finger tight, the spanner (wrench) should be used to give the nut only an extra half turn. Excessive tightening causes distortion in the fitting.

Similar fittings can be used for pipes made from hard plastic, such as HDPE or nylon. The edges of the brass ferrule bite into the surface of the plastic, giving a good grip. Softer plastics distort too easily and the ferrule can be pulled off the pipe surface. Brass and copper are expensive materials to use, so these types of pipe fitting are used much less than they have been in the past.

Many different types of compression fitting have been developed that are made from plastic. The sealing element, the equivalent of the ferrule, can be made from rubber or spring metal. The sealing element is held between a fitting with a suitably shaped end and a compression nut, shaped so that the element is compressed onto the pipe. Different systems are designed to suit different types of plastic pipe: the flexibility of the plastic material determines how effectively the compression element will seal against it.

The issue of cleanliness is even more important for these types of fitting. The pipe surfaces must also be very smooth, to give a surface against which the compression element can form a seal. A scratch in the pipe surface can form a path through which gas can leak. These fittings can be used with thin layers of high viscosity grease or flexible sealants. Excessive use of such sealants can affect the compression of the sealing element; only a thin layer should be used. These sealants can fill in very small scratches and imperfections in the surfaces of the pipe and fitting, but cannot compensate for larger scratches or distortions.

Push fittings

Push fittings are often used to adapt a metal pipe system to a plastic pipe, especially where the plastic pipe can act as a flexible connector between a metal gas appliance and a fixed gas line. The plastic pipe is pushed over a spigot, which often has a series of ridges that grip the inside of the pipe (see Figure A4.3). A gas-tight connection relies on the pipe being stretched, so the pipe's natural resilience causes it to grip the spigot. The pipe can be gently heated (for example in an oil bath) to soften it and allow it to stretch more easily. Most plastics contract as they cool, so the pipe fits over the spigot more tightly.

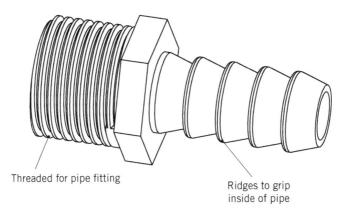

Threaded for pipe fitting

Ridges to grip inside of pipe

Figure A4.3 Diagram of a spigot for push fitting

Plastic or metal clips, such as a Jubilee clip (see Figure A4.4), are often used to hold the pipe more firmly on the spigot. However, such clips are not effective if the inside diameter of the plastic pipe is larger than the outside diameter of the spigot. When the clip is tightened, the plastic pipe becomes distorted and can form 'rucks' that allow gas to leak out. If the inside pipe diameter is slightly smaller than the spigot, so that the plastic is stretched, then the clip reinforces the tension supplied by the stretched pipe.

This type of fitting is usually used with less-dense polyethylene plastic pipes, which are more flexible and stretch more easily. Pipes made from rubber can also be used, especially to connect rigid pipes to gas appliances. Both plastic and rubber can adhere to the metal of the spigot, if left in place for more than a few days. If the pipe needs to be removed from the fitting, grease

Screw to tighten clamp

Variable sized ring

Figure A4.4 Diagram of a type of pipe clamp – Jubilee clip

can be used between the pipes and the spigot. If grease is used as a sealant, it is recommended that a clip be used.

Some piping systems use push fittings with HDPE and other harder types of plastic pipe, which must be warmed until the end of the pipe becomes soft enough to stretch over the fitting. The fittings can be made of metal or plastic.

Types of fitting

The main types of fitting that are usually available include straight connectors, 'T' pieces to allow pipes to include a branch connection, 'Y' pieces to allow one pipe to split into two and adapters to connect different sizes of pipe together. Pipe systems that use welded fittings often include adapters that allow connection to screwed fittings. There are also bulkhead fittings that allow a pipe to pass through a wall. There are two main types: a bulkhead fitting that opens into the space at one side of the wall, such as the inside of a tank, and a fitting which has pipe connections on both sides, so the pipeline continues across a wall. There is often a need to seal around the outside of a bulkhead fitting to prevent leakages.

Laying a pipeline

A biogas pipeline can be placed above or below ground. It can be made of steel or plastic pipe and connected together by fittings. Some plastics (such as PVC or PPR) have a short lifetime if used above ground, because the ultraviolet light in sunlight causes the plastic to depolymerize and it becomes steadily weaker. If used above ground, these plastics need to be protected from the sun. Other plastics (such as HDPE) are much less affected.

A main gas valve is usually placed close to where the pipe comes from the plant, so the pipeline can be closed off, if necessary. The pipe leads from this valve to the place where the gas is used, usually a kitchen. In some cases, there may be a splitter that allows the pipe to be laid to two different kitchens. The pipe is usually taken through the wall of the kitchen and the end fixed in a suitable place, with one or more spigots in easily accessible positions. Flexible hose is used to connect one or more gas burners to these spigots. A gas tap is provided for each gas burner, but this tap can either be mounted before the spigot on the wall, or be part of the biogas burner.

If metal pipe is used, the sections are screwed together with straight connecting joints between the two points. Plastic pipe usually comes in long reels, which can be unwound, so pipe can be cut to a suitable length. Connections need to be made between the main gas valve and the pipe at one end and the other end of the pipe and the spigots that supply the biogas to the kitchen. If the pipe is laid above ground, suitable supports are required that hold the pipe in the right place. The mounting points are usually fixed to poles so people can walk under the pipe. If the pipe is laid underground, trenches need to be dug, into which the pipe can be placed. If plastic pipe is used underground, the pipe should be laid in a thick layer of sand, to prevent rats and other rodents biting the pipe. This also prevents sharp stones in the soil puncturing the pipe.

The main issue with laying a biogas pipeline is that moisture can condense in the pipe and the water that forms can accumulate at low points and block the pipe. A biogas pipeline must slope smoothly to one or more low points from where the water can be removed by a variety of means. The slope should be at least 1:100 (Fulford, 1988), although it can be much steeper. One system for removing condensed moisture in an above ground system is to slope the pipe steeply upwards from the main gas valve (see Figure A4.5). If water condenses

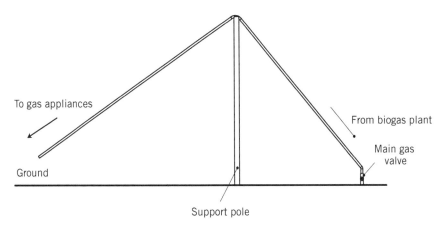

To gas appliances

Ground

From biogas plant

Main gas valve

Support pole

Figure A4.5 Diagram of above ground gas pipe with high point

in this section of the pipe, it automatically runs back into the digester pit. The pipe will pass over the top of a pole and then slope downwards towards the kitchen.

A longer pipe might need a series of poles, but the danger is that the pipe will form a series of loops, with a low point between each pole (see Figure A4.6). If such an arrangement has to be used, the user needs to drain the loops at least once a week. They should lift each loop successively, from the one furthest from the digester, to drain any water into the next loop. The water in the last loop can then be drained back into the digester pit. The user can check the pipe for cracks or other damage at the same time. A better method would be to arrange a series of poles with mounting points at differing heights, so the pipe slopes upwards from the main gas valve and downwards to the kitchen.

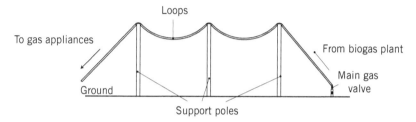

Figure A4.6 Diagram of above ground gas pipe with loops

If the pipe is laid underground, it needs to slope by at least 1:100 towards one or more water traps. The simplest type of water trap is a 'T' piece placed at the lowest point, with a short length of pipe connected to a tap (see Figure A4.7), placed in a brick-lined pit in the ground (Fulford, 1988). The water collects in the pipe and the tap can be opened to release the water regularly (e.g. once a week, depending on how much water condenses out). If the pipe has to run at a greater depth, a 'U'-tube trap can be made at the lowest point of the pipeline, with a small bucket on the end of a rod to collect and remove the water (Bulmer *et al.*, 1985).

A self-emptying water trap can be made by placing the end of the 'T' piece at the lowest point of the pipe in a water-filled container, such as a plastic bottle. The height of the water level in the container must be higher than the pressure of gas in the pipe, otherwise gas will bubble out through the water. As condensed water collects in the pipe, it drains into the container. Excess water will come out of the top of the container, so it must be allowed to drain away. The problem with this approach is that water can evaporate from the container faster than it collects. Once the water level has dropped to below the gas pressure, gas will be lost from the pipe. Another version is a 'U' trap that is filled with water and open at one end (see Figure A4.7). The arms of the 'U must be long enough so that the pressure supplied by the water in the open arm is higher than the maximum pressure from the biogas plant.

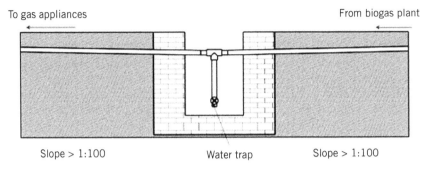

Figure A4.7 Diagram of underground gas pipe with water trap

Leak testing in gas pipelines

Once a pipeline has been connected up, but before it is made less accessible (such as being buried, if it is underground, or raised up on poles, if it is above ground), it needs to be tested for leaks. One way to do this is to connect a manometer to one of the spigots used for connecting gas burners, increase the gas pressure in the pipe and then close it off. If the pressure, as measured by the manometer, drops quickly, there is a leak in the pipe. If the pressure remains at the same reading, there are no leaks.

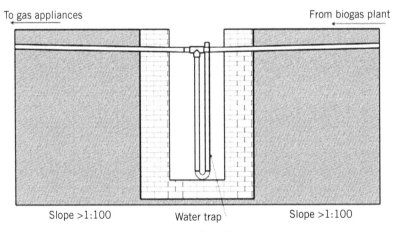

Figure A4.8 Diagram of underground gas pipe with self emptying water trap

A manometer can be made from two transparent tubes, connected by a 180° bend (see Figure A4.9). The straight sections should be longer than the maximum pressure in the biogas plant, as measured in mm of water. A transparent flexible plastic pipe can be mounted on a wooden board, held down by clips. The 'U'-shaped pipe is half filled with water. A water-soluble dye or ink can be used to colour the water. The straight sections of the pipe can have dimensions marked next to them, so the pressure can be read off. Such a manometer can be installed with a fixed-dome plant as an extra appliance. The manometer can be connected with a flexible pipe to one of the spigots at the end of the gas line, in the kitchen, so the cook can monitor the amount of gas left in the plant.

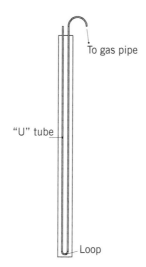

To gas pipe

"U" tube

Loop

Figure A4.9 Diagram of a simple manometer

If the gas pipe is to be checked for leaks before the plant is started, the gas valve to the plant is closed. If there is only one spigot in the kitchen, a flexible pipe can be used to connect it to a 'T' piece. One arm of the 'T' can be connected by a flexible pipe to the manometer and the other to a device used to pressurize the pipe, such as a bicycle pump. A valve is also required, so the pipe can be closed, once the manometer shows a high pressure.

If the manometer shows that there is a leak, it can be detected using soapy water. A mixture of detergent (such as soap) and water is made that forms sustainable bubbles. If this mixture is spread over a pipe connection, the bubbles that are generated reveal a leak. Once the connection has been remade, the pipeline must be checked for leaks again.

Once the plant has started generating biogas, the pipeline can be checked for leaks again. The gas from the plant can be used to pressurize the plant and the main gas valve closed off. The pressure can be checked with the manometer, to see if it drops or remains stable. The positions of any leaks can again be checked with soapy water.

Pressure loss down pipes

Fluids are driven through pipes by a pressure difference between the ends of the pipe. The pressure in a biogas plant depends on the way the gas is stored. The weight of the floating drum generates pressure in that type of plant. The difference in the height of the liquid surfaces between the slurry reservoir and the main digester tank provides the pressure from an underground dome plant. The pressure in a flexible-bag plant is provided by a weight placed on the bag. As gas flows down the pipe, pressure is lost through friction. As the flow rate increases, the pressure loss increases. If the pressure losses are greater than the driving pressure, the gas will not flow down the pipe.

Pressure losses are affected by the internal diameter of the pipe and its length, as well as the gas flow rate. The flow can be one of two types, laminar and turbulent, which also affects the pressure loss. The pressure loss can be calculated from measurements of the pipeline (Douglas *et al.*, 2000).

The flow velocity of the gas (v) depends on the volume flow of the gas (Q) and the pipe diameter (d): $v = \dfrac{4\,Q}{\pi\,d^2}$ If Q is measured in m^3 s^{-1} and d is measured in m, the velocity v is in m s^{-1}.

The gas used by various appliances is given in Table A4.2.

The pressure losses in a pipe depends on the value of the fiction factor: f, which depends on the Reynolds number: Re.

$\text{Re} = \dfrac{\rho\,d\,u}{\mu} = \dfrac{d\,u}{v}$ where ρ_w=biogas density ($=1.0994$ kg m^{-3}) and μ=biogas viscosity ($\dot= 1.297 \times 10^{-5}$ kg m^{-1} s^{-1}). $v = \dfrac{\mu}{\rho}$ is the kinematic viscosity of biogas ($=1.797 \times 10^{-5}$ Pa s).

Table A4.2 Gas consumption of various appliances

	Biogas consumption (75 mm WG)	
Biogas appliance	m^3 min^{-1}	m^3 hr^{-1}
Gas lamp (per mantle)	0.0024	0.14
Refrigerator burner	0.0024	0.14
Domestic burners (stoves)	0.004 to 0.015	0.2 to 0.9
Commercial burners	0.02 to 0.05	1 to 3
Dual fuel engines (per kW)	0.009	0.56
Spark engines (per kW)	0.012	0.7

The equation used for the friction factor (f) depends on the value of Re.

For laminar flow (which is most likely for the flow of biogas in pipes), Re < 2000 and $f = \dfrac{64}{\text{Re}}$. If Re > 4000, the flow is turbulent and $f = \dfrac{0.316}{\sqrt[4]{\text{Re}}}$. If Re is between 2000 and 4000, the flow is transitional and the value of f is somewhere between the values given by the equations.

The pressure drop (Δp) can then be calculated:

$$\Delta p = f \frac{L}{d} \frac{\rho \, v^2}{2},$$ where L is the length of the pipe in m, and Δp is measured in Pa (N m^{-2}).

The pressure loss can be measured in m of water (h), using $\Delta p = h \, \rho_w \, g$, where ρ_w is the density of water (1000 kg m^{-3}) and g is the acceleration due to gravity = 9.81 m s^{-2}.

Example A4.1 Pressure loss along a pipe

Two domestic burners are supplied through a 15 mm diameter pipe of length 50 m.

Taking each burner as using 0.5 m^3 per hour, the total flow $Q = 1$ m^3 hr^{-1} or 0.00028 m^3 s^{-1}.

$$u = \frac{4 \times 0.00028}{\pi \times 0.015^2} = 1.57 \text{ m s}^{-1}.$$ The Reynolds Number: $= 1310$ $\text{Re} = \dfrac{0.015 \times 1.57}{1.797 \times 10^{-5}}$

The flow is laminar (Re < 2000), so $f = \dfrac{64}{1310} = 0.0488$ and

$$\Delta p = 0.0488 \times \frac{50}{0.015} \times \frac{1.0994 \times 1.57^2}{2} = 221 \text{ Pa, which is 22.5 mm water}$$
gauge.

This is a high pressure-drop for gas from a floating drum plant, so a larger pipe size should be used.

If the gas is from a fixed-dome plant (GGC or Deenbandhu), this pressure drop is acceptable.

References

Bulmer, A., Finlay, J., Fulford, D.J. and Lau-Wong, M.M. (1985) *Biogas: Challenges and experience from Nepal Vols. I and II*, Butwal, Nepal: United Mission to Nepal. <www.kingdombio.com/Biogas-vol-I.pdf> [accessed 23 July 2014].

Douglas, J.F., Gasiorek, J. and Swaffield, J. (2000) *Fluid Mechanics*, 4th edn, Prentice Hall. Harlow, UK<http://bib.tiera.ru/b/104247> [accessed 23 July 2014].

Fulford, D. (1988) *Running a Biogas Programme: A handbook*, London: Practical Action Publications (Intermediate Technology Publications). <http://developmentbookshop.com/running-a-biogas-programme-pb> [accessed 23 July 2014].

The Engineering Toolbox (2010) 'Carbon, Alloy and Stainless Steel Pipes-ASME/ANSI B36.10/19' [website-information] The Engineering Toolbox. <www.engineeringtoolbox.com/asme-steel-pipes-sizes-d_42.html> [accessed 23 July 2014].

APPENDIX 5
Gas appliance design

Abstract

The design of biogas burners depends on the properties of biogas. The basic theory for the design of biogas burners uses fluid-flow theory and defines the sizes of key parameters, including the diameters of the gas jet and the mixing-tube throat. The total area of the flame ports is another important dimension. Methods can be used to improve the stability of gas flames.

Keywords: biogas, anaerobic digestion, chemistry, thermodynamics

Biogas properties

The properties of biogas are given in Table A5.1. The lower heating value (enthalpy of combustion) assumes that the water vapour generated by the reaction remains in the vapour phase. The range of values relates to the possible ratios of methane to carbon dioxide in biogas. Various properties (such as density) are affected by temperature and pressure, so the values at other temperatures and pressures can be adjusted using the perfect gas law: $p = \rho R T$, where p is pressure, ρ is density, T is absolute temperature and R is the gas constant.

Table A5.1 Properties of biogas (Fulford, 1988)

Biogas: assumed 58% CH_4 and 42% CO_2, saturated with water vapour at 30 °C and standard pressure (1 bar)		
Enthalpy of combustion (H)	21.5 MJ m^{-3}	20.1 to 25.9 MJ m^{-3}
Effective molecular weight	27.35	24 to 29
Density (ρ)	1.0994 kg m^{-3}	0.96 to 1.17 kg m^{-3}
Specific gravity (sg)	0.94	0.82 to 1.00
Viscosity (μ)	12.97 × 10^{-6} kg m s^{-1}	
Optimum air to fuel ratio	5.52:1	15% biogas
Flammability limits	9% to 17% biogas in air	
Wobbe number (W)	22.2 MJ m^{-3}	
Burning velocity	0.25 m s^{-1} in air	

The lower heating value (21.5 MJ m^{-3}) is less than for natural gas (48.5 MJ m^{-3}), which is the main reason that gas appliances designed for natural gas cannot be used for biogas. The optimum air to fuel ratio (5.5:1)

also affects the way the burner is designed. The Wobbe number ($W = \dfrac{H}{\sqrt{sg}}$) is a measure of the way different fuel gases perform in burners. The value for biogas (22.2 MJ m^{-3}) is very different from that for natural gas (53.7 MJ m^{-3}) and LPG (86.8 MJ m^{-3}), which shows that burners need to be designed specifically for biogas.

Gas burners

A biogas burner mixes a fuel gas with air and supplies the mixture to flame ports, where it burns in air. The amount of air required to burn biogas can be calculated from the combustion equation:

$$0.58\,CH_4 + 0.42\,CO_2 + 1.16\,O_2 + 4.36\,N_2 \rightarrow 1.0\,CO_2 + 1.16\,H_2O + 4.36\,N_2$$

The nitrogen is not involved in the combustion process, but is an essential component of the air (occupying 71 per cent of the mixture). The exact amount of air required to burn the fuel completely is called the stochiometric air requirement. Balancing the above equation shows that 1 part biogas requires 5.52 parts of air to burn completely. In a burner the required amount of air is provided in two parts. Part of the required air, the primary air, is mixed with the biogas inside the burner. When the mixture is supplied to the flame ports the gas burns in the secondary air, which comes from the outside of the burner.

Burners are often designed so the primary airflow can be controlled. Reducing the primary airflow means the gas needs to find extra air from the outside of the burner, so it burns with a lazy flame. Increasing the primary airflow reduces the height of the flame at the flame ports, but there is a risk of flashback, where the flame jumps through the burner ports and burns within the mixing tube. Biogas is mixed with primary air in the correct proportions in a mixing tube open to the air (see Figure A5.1). The mixing tube leads to the burner ports and should have a length that is at least ten times the diameter to ensure good mixing of biogas and air so they burn evenly.

The theory behind the design of a biogas burner is based on fluid dynamics (Chandra *et al.*, 1991). This theory has been incorporated into key equations

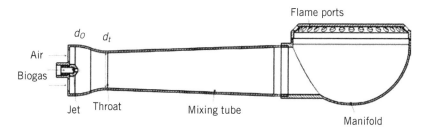

Figure A5.1 Basic parts of a biogas burner

that can be used in burner design (Pritchard *et al.*, 1977). The gas flow is controlled by the gas jet, a small nozzle with a hole in it. The volume flow rate (Q) is controlled by the area of the hole (A_0, m²) (Fulford, 1988):

$$Q = 3.16\, C_d\, A_0 \sqrt{\frac{p}{s}} \ \ \text{m}^3\ \text{s}^{-1},$$ where p is the gas pressure (Pa) and s the specific gravity of the gas. C_d is the coefficient of discharge of the jet (with a value between 0.75 and 0.95, depending on how well the jet is made). The gas flow also determines the power of the burner (P, kW), the amount of heat delivered per second as $P = Q \times H$, so:

$$P = 3.16\, H\, C_d\, A_0 \sqrt{\frac{p}{s}} \times 1000 \ \ \text{kW} \ \ \text{or} \ \ P = 3.16\, W\, C_d\, A_0 \sqrt{p} \times 1000 \ \ \text{kW},$$ using the Wobbe number.

Air is drawn into the tube by the venturi effect as the gas flow reduces in velocity and therefore in pressure as it expands from the jet into the tube. The entrainment ratio, (r) the amount of air that is drawn in by the pressure reduction in the gas flow, is given by Prigg's formula:

$$r = \sqrt{s}\left(\sqrt{\frac{A_t}{A_0}} - 1\right) \ \ \text{or} \ \ r = \sqrt{s}\left(\frac{d_t}{d_0} - 1\right),$$ where A_t is the throat, the narrowest part of the mixing tube and d_0 and d_t are the diameters of the jet and the throat, respectively. The entrainment (r) is usually chosen to be between 50 per cent and 75 per cent of the stochiometric value (5.5). The mixing tube can be either shaped as a venturi, tapering towards the throat and away from it, or as a straight pipe. The length of a straight mixing tube should be at least $10 \times d_t$, while a venturi mixing tube can be shorter: $6 \times d_t$.

The other end of the mixing tube leads to the manifold, in which the mixture is directed to the flame ports. As the mixture comes from the flame ports, it burns in the secondary air that flows from below the ports. A key parameter that determines the size of the flame ports is the flame velocity. If the flow velocity of the mixture from the flame ports is higher than the flame velocity, the flame lifts off the flame ports and goes out.

$$v_f = \frac{(r+1)\,Q}{A_p} < 0.25 \ \text{m s}^{-1}, \ \text{or} \ A_p > \frac{(r+1)\,Q}{0.25} \ \text{m}^2$$

The burner ports are often arranged in a circle; most pots have a circular base which needs to be evenly heated. Free circulation of the 'secondary' air below and above the burner ports is required, to allow full combustion to take place, as well as to allow combustion products (carbon dioxide and water vapour) to easily escape. The air flowing towards one flame port should not interfere with the air flowing towards another port.

The distance between the flames from the burner ports and the pot is important. If the pot is too high, too much heat is lost. If the pot is too low, it can 'quench' the flames by cooling them, and prevent complete combustion. The ideal position for the pot is just above the tip of the flame (see Figure A5.2). The height of the flame can be altered by adjusting the airflow into the burner. Less air usually gives longer flames and increasing the primary air supply means the flames are shorter.

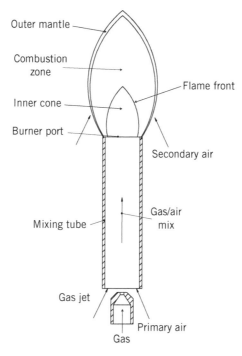

Figure A5.2 Shape of a gas flame

The DCS stove supported the mixing tube and burner port assembly in a metal frame into which a cast iron 'grate' was placed to support pots. Since both flat- and round-bottom pots were often used in Nepal, the grate included supports for both types of pot, depending on which way up it was placed in the frame. The main frame must be strong and stable, so the burner does not wobble when food is being stirred during cooking.

Example A5.1 Design of a biogas stove

The DCS burner was designed to deliver 5 kW, a typical power for a domestic stove, from a gas pressure of 740 Pa (75 mm water gauge).

$$Q = \frac{P}{H} = \frac{5}{21.5 \times 1000} = 2.33 \times 10^{-4} \text{ m}^3 \text{ s}^{-1} \text{ so } A_0 = \frac{2.33 \times 10^{-4}}{3.16 \times 0.8 \times \sqrt{740/.94}} = 3.29 \times 10^{-6} \text{ m}^2.$$

This gives a hole size for the jet of 2.05 mm. A standard size would be 2 mm. Using an entrainment ratio of 53% ($r=2.93$), gives a value of $d_t = \left(\dfrac{2.93}{\sqrt{0.94}} + 1 \right) \times 2 = 8.04$ mm (assume 8 mm).

If the mixing pipe is straight, the length should be at least 80 mm.

Using the flow velocity equation: $A_p > \dfrac{3.92 \times 2.33 \times 10^{-4}}{0.25} > 3.64 \times 10^{-3} \text{ m}^2$, to give a stable flame.

If each flame hole has a diameter of 5 mm $A_h = \dfrac{\pi \times \left(5 \times 10^{-3}\right)^2}{4} = 1.96 \times 10^{-5} \text{ m}^2$, the number of flame holes should be at least 185, to get good flame combustion.

The DCS burner was designed so the burner ports are drilled in a removable cap over a manifold into which the biogas/air mix is fed. This cap can be taken off and the burner ports cleaned, in case food and grease are spilled over them. However, the original DCS stove was designed with an inadequate number of flame ports, so the efficiency was lower than it should have been. Better gas burners have a burner manifold made in the shape of a donut, so that flame ports can be drilled in the inner surface of the donut as well as the outer one (see Figure A5.3).

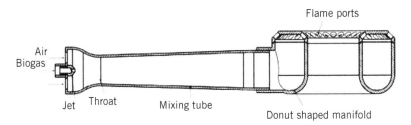

Figure A5.3 Improved biogas burner

Even if the total flame port area is too small, flame stability can be improved (Fulford, 1988) by various methods (see Figure A5.4). A ledge below the flame allows a greater flow of secondary air. Flame ports can be made on the side of the manifold, which also gives better secondary airflow. Small air ports (metering orifices) can be drilled around the outside of a main flame port, which provides for extra small flames that have the effect of holding the main flame onto the manifold. Flame ports do not need to be round. Rectangular slits around the side of a cylindrical manifold allow air in from the sides of the flame as well as from underneath.

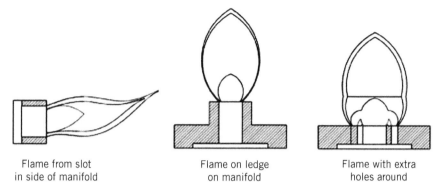

| Flame from slot in side of manifold | Flame on ledge on manifold | Flame with extra holes around |

Figure A5.4 Ways to stabilize flames

References

Chandra, A., Tiwari, G.N. and Yadav, Y.P. (1991) 'Hydrodynamical modelling of a biogas burner', Energy Conversion and Management 32: 395–401 <http://dx.doi.org/10.1016/0196-8904(91)90058-Q> [accessed 17 July 2014].

Fulford, D. (1988) Running a Biogas Programme: A handbook, London: Practical Action Publications (Intermediate Technology Publications). <http://developmentbookshop.com/running-a-biogas-programme-pb> [accessed 23 July 2014].

Pritchard, R., Guy, J.J. and Connor, N.E. (1977) Industrial Gas Utilization: Engineering principles and practice, Epping UK: Bowker (for British Gas). <http://books.google.co.uk/books/about/Handbook_of_industrial_gas_utilization.html?id=iPhSAAAAMAAJ&redir_esc=y> [accessed 17 July 2014].

APPENDIX 6
Follow-up surveys

Abstract

Surveys are part of the work of follow-up, which is a key part of a biogas extension programme. Surveys start with secondary sources: existing information on the survey area. Different types of survey include inspection, questionnaires, group interviews, use of key respondents and field measurements. Surveys must be carefully planned and then tested by using a pilot survey.

Keywords: biogas, anaerobic digestion, chemistry, thermodynamics

Value of follow-up surveys

One aspect of the original Development and Consulting Services (DCS) programme in Nepal was the use of follow-up surveys. This was incorporated into the approach used by the Gobar Gas Company (GGC) (Fulford *et al.*, 1991) and then the Biogas Sector Partnership of Nepal (BSP-N) (Pradhan *et al.*, 2007). It is seen as an essential part of the biogas programmes started by Netherlands Development Organization (SNV) in other countries, such as Bangladesh (iDE, 2011) and Vietnam (AMDI, 2012). Surveys identify a wide range of issues that laboratory and test-site research work is unable to highlight; biogas plant owners will use their systems in ways that are often not predicted by experts. Development protocols that have been defined as an effective approach to biogas extension may not work as expected. Surveys generate data that can be used to adapt programme plans.

The type of information generated by surveys can be in the technical, development and financial areas of a programme. However, a survey requires time and financial resources, so should usually be done by a team of people with a range of expertise, so that different subject areas can be covered. Another requirement is that a large number of people visiting a customer can be intimidating, so a small team is more effective. The survey approach can be planned using advice from a larger team of experts, but the actual survey requires a much smaller group of people.

Most carbon-offset contracts specify the requirement for follow-up surveys, as the regular payment of finance is dependent on projects continuing to replace fossil carbon. The programme must be able to demonstrate that biogas plants continue to be used and are fulfilling the expectations defined in the contract. Carbon-offset finance companies require a detailed audit of programmes, with careful checks between proposed and actual outcomes (NESS, 2011).

There are many publications and courses on the subject (Cohen, 2009; Crawford, 1997), so this appendix offers an overview of how it is applied to biogas extension programmes.

Secondary sources of information

A survey is expensive in terms of resources: time, personnel, and finance, so it is important that there is no duplication of work. The first stage of any survey is the desk study. Any previous information that may relate to the subject of the survey should be obtained and studied.

For biogas follow-up surveys the main secondary source of information is the records that the project already holds on the biogas plants that have been built. For a wider marketing survey, such as when the programme plans to start extension work in a new geographical area, secondary sources may include census data, records of surveys done in the area by other groups, such as agricultural extension workers or groups doing extension work for other renewable energy technologies, such as solar. Discussions with field staff from the Department of Agriculture, who are working in the area, will provide useful background information. Other useful sources of information may come from Micro-Finance Institutions (MFIs) and agricultural development banks working in the area.

Types of survey

The approach to a survey depends on the type of information that is required, although a single survey may adopt several approaches to allow the collection of various types of information.

Inspection

The simplest approach is to 'look and see'. The first step in a biogas follow-up survey is to visit an address and check whether the number on the biogas plant matches the one in the records. The second step is to check that the biogas plant is producing gas and that the gas burners are working effectively. There is a series of further inspections that can be made, such as the number of animals, if the plant is fed with animal dung, or the weight of daily feed material. When the survey is planned, a checklist can be defined, so that each item can be ticked off.

Questionnaires

Biogas plant users, and non-users, can be asked a set of defined questions to provide required information. The opinions of a sample of users can provide an idea of the effectiveness of a programme. The required information may range from the performance of the installers, to the overall usefulness of the technology. Since each person is asked the same set of questions, it is possible to calculate an average opinion from the sample.

In order to aid analysis, the questions can be made to be multi-choice; a grading system can be offered as to whether the respondent considers something to be good or bad. A grading scale can be used, with respondents asked to assign a number to their opinion. It is better to offer an even number of possible options (1 to 6, rather than 1 to 5) so people are not tempted to choose the middle value. The range of possible values should usually be 10 or less (1 being very bad and 10 being very good). When the survey is analyzed, the numbers can easily be entered into a spreadsheet, so that averages and standard deviations can be calculated.

The design of a questionnaire must be done carefully. The questions need to be written with the idea in mind that the results need to be analysed. Questions should be 'open': they should not be written as if a particular answer is expected. They also need to be unambiguous. The simplest and most obvious questions are the best, such as 'What do you see as the benefits of using a biogas plant?'. It is important that respondents are not made to feel they are sitting an examination with right and wrong answers. The questioner needs to adopt an attitude that the respondents' opinions are required and they are important for the survey.

Discussions

Some of the questions in a questionnaire can be designed to elicit people's own ideas. These answers cannot be analyzed in a spreadsheet, but can provide deeper insights into people's wider thinking. Again, it is often the simplest and most obvious questions that produce such answers. People will offer an obvious answer, but will then go on to explain their thinking, revealing unexpected ideas.

This type of information can also be obtained through group discussions. In many places, if one or two people are talking in a public place, many will come to join the conversation. The opinions that people offer will be those of the group, assuming that everyone agrees with the points made. If there is a disagreement over certain answers, a discussion about the issue can reveal even more information. This approach is a valuable tool for use in a marketing survey. The use of focus groups – a group of people chosen to reflect local opinions – is a version of this approach. Discussions can be held with the focus groups at different times, to reveal how local opinions change as a programme progresses.

Interviews with key respondents

Key respondents are people who have insights into local attitudes and opinions. For a general marketing survey, such people may include teachers, community leaders, and faith leaders, who have regular contact with a group of local people and have an understanding of their thinking. For follow-up surveys, key respondents will include staff of the MFIs and banks involved in providing loans to customers and agricultural extension workers, who are regularly visiting local farmers and understand the impact of the provision of biogas plants. Other key respondents are the staff of the extension programme, who are responsible for selling biogas plants, building them, or supervising the building work.

Measurements

Questionnaires and interviews provide qualitative information, based on the opinions of people. Quantitative information must be obtained by measuring parameters, such as gas production and parameters of the feed materials used.

The key follow-up visit should be made once a plant has been built, but before it is filled with feed material. An assessor can then take measurements inside the plant and check that it has been built according to the plans provided. BSP-N set up a detailed check list for the quality of the plants that were built under their programme (Lam, 1996).

The performance of biogas plants in the field can be evaluated by taking measurements of the input and output from the plant. The mass of the feed material can be weighed before it is added. Samples of this feed material can be taken, so that its properties (TS, VS, COD, biogas potential, etc.) can be tested in a laboratory (see Chapter 4). The effluent can be collected in containers and weighed. Again samples can be taken so that its properties can be tested. The total volume of the gas produced can be measured with a meter. This work is very time consuming, so such measurements are only done occasionally. Carbon-offset finance projects do require this type of data (Devkota, 2003), so they can assess the value of the gas produced (Zifu *et al.*, 2009).

Planning a survey

The key to planning a survey is to define clearly what information is required. The survey methods to be used and the questions used are directly defined by the answers that are needed. A survey that is designed to identify weaknesses in the extension programme, such as in the technical designs and management approach, will be different from one that is required by a carbon-offset finance programme. A good survey is designed around a tight focus on the information that is being sought.

Surveys require resources, such as personnel and time, so need to be carefully planned to make the best use of them. If there is a large number of possible data points, such as the number of plants that have been built, a sample needs to be selected. If one of the purposes of the survey is to compare different aspects of the work, such as the influence of working in different geographical zones within an area, the sample should include plants that reflect these different aspects. Otherwise, a sample should be randomized, each 'data point', such as each customer, should have an equal chance of being selected.

If the list of customers is already on a database, which is recommended (see Chapter 10), then it is fairly easy to select a random sample using suitable software. A large sample gives more reliable statistical results, but demands more resources. Most surveys will use a range of different types of survey methods, although the majority will include the use of questionnaires.

It is useful to carry out a pilot survey on a limited number of customers. The planned approach and especially the questionnaire can be tested. If respondents understand certain questions poorly, then the questions may need to be rewritten. If detailed measurements prove more difficult to make than was expected, the aims of the survey may need to be re-evaluated. The pilot survey also enables the personnel involved in the work to be trained in the desired approach.

Analyzing and reporting

The survey design needs to take into account how the data will be analyzed. A copy of the database can be used to record the data that is collected. The data can then be analyzed by transferring it to a spreadsheet and using the statistical functions available. Most data will be recorded on paper, as this is the most flexible medium. It can be supplemented by records on electronic media, especially those that can record GPS data, so that locations can be defined accurately.

The first step in analyzing the data is to enter it into suitable software, such as a spreadsheet. The survey design should include the software that will be used to enter and analyze the data. Time and other resources need to be allowed for the work of data entry. Depending on the information that is required, the analysis of the data may require statistical analysis and suitable software that allows this to be done.

There should be a place in the process of data entry to allow for unexpected responses. Often new information can be obtained from people who provide answers that do not fit into the defined framework.

References

Asian Management and Development Institute (AMDI) (2012) Biogas User Survey, Vietnam 2012, Hanoi, Vietnam: SNV and MARD. <www.snvworld. org/en/download/publications/biogas_user_survey_vietnam_2012.pdf> [accessed 23 July 2014].
Cohen, A. (2009) The Multidimensional Poverty Assessment Tool: Design, development and application of a new framework for measuring rural poverty, IFAD, International Fund for Agricultural Development. <www.ifad.org/mpat> [accessed 23 July 2014].
Crawford, I.M. (1997) Marketing Research and Information Systems, FAO Food and Agriculture Organisation. Rome <www.fao.org/docrep/w3241e/ w3241e00.htm#Contents> [accessed 23 July 2014].
Devkota, G.P. (2003) Analysis of Biogas Leakages from Household Digesters, Kathmandu: Winrock International.

Fulford, D., Poudal, T.R. and Roque, J. (1991) Evaluation of On-going Project: Financing and construction of biogas plants, United Nations Capital Development Fund. New York

iDE (2011) Biogas user survey, Bangladesh 2010, IDCOL Dacca, Bangladesh. <www.snvworld.org/en/download/publications/biogas_user_survey_2010_ bangladesh_2011.pdf> [accessed 17 July 2014].

Lam, J. (1996) Enforcement of Quality Standards up on Biogas Plants, Kathmandu, Nepal: BSP (Biogas Support Programme), SNV. <www.snvworld. org/en/download/publications/enforcement_quality_standards_ nepal_1996.pdf> [accessed 17 July 2014].

NESS, N.E. and S.S. (2011) Biogas User Survey, Nepal 2009–2010 (CDM Activity 1), AEPC. <www.snvworld.org/en/download/publications/biogas_user_ survey_2009-2010_cdm_activity_1_nepal.pdf> [accessed 17 July 2014].

Pradhan, S., Shrestha, M., Shrestha, M.L., Sharma, I., Sitoula, M.L. and Kharel, H.R. (2007) Biogas User Survey, Nepal 2007, BSP-N (Biogas Sector Partnership-Nepal) Kathmandu, Nepal. <www.snvworld.org/en/download/ publications/biogas_users_survey_2006-07_final_report_nepal.pdf> [accessed 17 July 2014].

Zifu, L., Mang, H.P. and Neupane, K. (2009) Biogas Audit: Summary of findings & action plan (volume I), KfW & USTB for SNV, Frankfurt, Germany. <www.snvworld.org/en/download/publications/biogas_audit_nepal_ volume_1_2009.pdf> [accessed 17 July 2014].

Index

Milton Keynes UK
Ingram Content Group UK Ltd.
UKHW011342220924
1771UKWH00072B/45